FinOps Handbook for Microsoft Azure

Empowering teams to optimize their Azure cloud spend with FinOps best practices

Maulik Soni

BIRMINGHAM—MUMBAI

FinOps Handbook for Microsoft Azure

Group Product Manager: Mohd Riyan Khan

Publishing Product Manager: Suwarna Rajput

Senior Editor: Sayali Pingale

Technical Editor: Arjun Varma

Copy Editor: Safis Editing

Associate Project Manager: Ashwin Kharwa

Proofreader: Safis Editing

Indexer: Manju Arasan

Production Designer: Shankar Kalbhor

Marketing Coordinator: Agnes D'souza

First published: May 2023

Production reference: 1200423

Published by Packt Publishing Ltd.
Livery Place
35 Livery Street
Birmingham
B3 2PB, UK.

ISBN 978-1-80181-016-6

www.packtpub.com

This book would not have been possible without the unwavering support and encouragement from the following people. My mother, your love and guidance have been the foundation upon which I have built my life. My father, your wisdom and strength have been an inspiration to me. My wife Ripal, your belief in me and your unyielding devotion have given me the courage to pursue my dreams. And my son Shiven, you are the reason I strive to be a better person every day.

– Maulik Soni

Contributors

About the author

Maulik Soni is a principal cloud solutions architect with a track record of helping commercial enterprise customers successfully adopt and implement Azure core infrastructure services. He has a bachelor's degree in advanced accounting and auditing and an advanced diploma in computer business applications from the Department of Electronics and Accreditation of Computer Classes in New Delhi.

Maulik has delivered more than a dozen cost optimization assessments, providing actionable recommendations that have helped Fortune 50 companies save between $1-3 million per year in cloud costs. When he's not at work, Maulik indulges his passion for street photography. He also enjoys exploring new destinations with his wife and son.

Special thanks to my friend and guide, Nirbhay Anand. Your unwavering support and encouragement throughout the writing of this book have been invaluable.

This book would not have been possible without the support from the entire Packt publishing team, specifically Suwarna Rajput and Niranjan Naikwadi for reaching out and convincing me to write this book, Ashwin Kharwa for constantly keeping me on the schedule, Sayali Pingale for coordinating smooth hand-offs during the complex editing process, and Agnes D'souza for wonderful coverage of graphics designs.

About the reviewers

Mustafa Mamawala has been working in the IT industry for more than 29 years, and he has over 16 years of experience in designing software systems and providing solutions as an architect. He has played key roles in digital transformation initiatives, cloud transformation journeys, monolith to microservices transformations, agile and DevOps automation, application/product architecture review, data analytics, and engineering management.

Mustafa has led various communities of practice to build capabilities in Microsoft technologies, various cloud services such as Azure and Google Cloud Platform, and content management systems across organizations he has worked.

He currently works as a lead enterprise architect with HSBC Software Development India, based out of Pune.

I would like to thank my wife Tasneem and son Ebrahim, who helped me manage my time and commitments.

I would also like to thank my associate Nirbhay Anand, who introduced me to Packt Publications, and through whom I got this opportunity.

Special thanks to my friend Jaywant Thorat (Microsoft Certified Trainer and PowerBI expert) for helping me overcome environmental challenges while reviewing the hands-on activities in this book.

Nirbhay Anand has more than 16 years of solid software design, development, and implementation experience. He has developed software in different domains such as investment banking, manufacturing, supply chain, power forecasting, and railroad contract management.

Being a Microsoft techie, he has expertise in Azure and cloud computing, and as a technical program manager, he has been engaged with FinOps management.

Currently, he is associated with CloudMoyo, a leading cloud and analytics partner for Microsoft. CloudMoyo brings together powerful BI capabilities using the Azure data platform to transform complex data into business insights.

He is a passionate blogger and book reviewer.

I would like to thank my wife Vijeta and kids Navya and Nitrika (Nikku) for their support. I would also like to thank my friends, family, and well-wishers for their never ending support.

Table of Contents

Preface   xiii

Part 1: Inform

1

Bringing Visibility and Allocating Cost   3

Technical requirements   4   What is ABC allocation?   13

Tools used in this book for   Cost allocation in Azure for FinOps   15
implementing FinOps for   Cost allocation using the account,
Microsoft Azure   4   management group, and
Azure CLI   4   subscriptions hierarchy   15
Power BI Desktop   4   Cost allocation using resources tags   17
Azure Cost Management + Billing   5
Azure Advisor   5   Exploring cost analysis in the
Azure Monitor   6   Azure portal   18
Azure Pricing Calculator   6   Identifying the offer type for your
                                                     subscription(s)   19
What is the Microsoft Azure   Accumulated and forecasted cost   20
Well-Architected Framework?   7   Cost grouped by service   21
Creating a baseline using the WAF   Cost grouped by management group   22
Cost Optimization assessment   7   Cost grouped by tag   24
Cost allocation from an accounting   Creating, saving, and sharing custom cost
point of view   13   analysis views   24

                                                     Summary   26

2

Benchmarking Current Spend and Establishing Budgets   27

Technical requirements	27	Production marketing website budget	34
The on-demand and elastic nature of Azure	28	Marketing development budget	37
		Marketing production budget	37
Developing KPIs for consistent reporting	28	Overall Marketing department budget	37
		Tracking the budget spend	38
Strategic and operational KPIs	28	**Creating and managing alerts in Azure cost analysis**	**38**
Leading and lagging KPIs	28		
Why do you need KPIs?	29	Budget alerts	38
Defining, measuring, and reporting KPIs	29	Spending anomaly alerts	40
Benchmarking between teams	31	Summary	42
Creating and managing budgets in Azure cost analysis	33		

3

Forecasting the Future Spend   43

Technical requirements	44	Forecasting based on past usage	51
Introduction to forecasting	44	Advanced forecasting by application	55
Getting your Azure usage data	45	Identifying usage charges by application	55
Setting up the Cost Management connector for Power BI	46	Fully loaded costs in forecasting	58
		Summary	61
Forecasting based on manual estimates	50		

4

Case Study – Beginning the Azure FinOps Journey   63

Case study – Peopledrift Healthcare	63	Benefits	66
Challenges	64	Summary	67
Objectives	64		
Solution	64		

Part 2: Optimize

5

Hitting the Goals for Usage Optimization   71

Technical requirements	72	Target 3 – enabling Azure Hybrid Benefit for Windows and Linux VMs	84
The project management triangle method for goal setting	72	Target 4 – right-sizing underutilized SQL databases	86
Setting OKRs or KPIs	73	Target 5 – enabling Azure Hybrid Benefit for SQL databases, managed instances, and SQL VMs	87
OKR examples	73		
KPI – tagging by business unit	74		
KPI – cost avoidance for unattached disks by business unit	74	Target 6 – upgrading storage accounts to General-purpose v2	88
KPI – Azure Hybrid Benefit utilization by business unit	75	Target 7 – deleting unattached discs	89
		Target 8 – deleting unattached public IPs	91
KPI – storage accounts with hot, cool, and archive tiers	76	Target 9 – Azure App Service – using the v3 plan with reservations and autoscaling	92
Understanding Azure Advisor recommendations for usage optimization	76	Target 10 – Azure Kubernetes Service – using the cluster autoscaler, Spot VMs, and start/stop features in AKS	93
Accessing Azure Advisor using the portal	77	**Trade-offs of cost versus security, performance, and reliability**	94
Accessing Azure Advisor using the CLI	78	Cost versus security	95
Top 10 usage optimization targets using custom Azure workbooks	80	Cost versus performance	95
		Cost versus reliability	95
Target 1 – 98% of all your resources must be tagged	80	**Summary**	96
Target 2 – right-sizing underutilized virtual machines	82		

6

Rate Optimization with Discounts and Reservations   97

Technical requirements	98	The Microsoft Enterprise Agreement	98
Commitment-based discounts in Azure	98	The Microsoft Azure Consumption Commitment (MACC)	99

Identifying reservation
opportunities for your workload   100

Using the Azure Cost Management
(ACM) Power BI app   100

Scenario 1 – you are purchasing VM
reservations for the first time   102

Scenario 2 – you have existing reservations
but want to purchase a new one for another
Region and VM SKU   104

Understanding Azure Advisor
recommendations for reservations   105

Reservation purchase and cadence   107

Purchase cadence   111

Reservation details, renewal,
savings, and chargeback report   112

Reservation details   112

Auto-renewal   114

Reservation savings and chargeback report   114

Reservation exchange and
cancellation   116

Exchange reservations   116

Cancel (return) a reservation   119

Summary   120

7

Leveraging Optimization Strategies   121

Technical requirements   122

Introducing Azure Spot market   122

Estimating Spot VM discounts   122

Spot VM and VM Scale Sets   123

Spot VM caveats   127

Eviction type and policy   127

Limitations   127

Pricing history and eviction
rate details   128

Architecting the workload to
handle eviction   130

Spot Priority Mix   132

Discounting strategies with
savings plans   133

Savings plan versus reserved instances   134

Purchasing a savings plan   135

Writing a business case for cost
optimization   136

Business case: Orion business analytics
platform cost optimization   136

Summary   139

8

Case Study - Realize Savings and Apply Optimizations   141

Case study – Peopledrift Inc., a
healthcare company   142

Challenges   142

The solution   142

Benefits   146

Summary   147

Part 3: Operate

9

Building a FinOps Culture 151

Technical requirements	152	Automated VM shutdown	
Establishing a CoE for cloud cost		and startup	156
management	152	Automated budget actions	162
Motivating engineering teams		Third-party FinOps tools	168
to take action	152	Apptio Cloudability	168
Incentivizing the team	153	CloudHealth by VMware	169
Penalizing the team	153	Cast.ai	169
Automated tag inheritance,		Summary	169
governance, and compliance	153		

10

Allocating Costs for Containers 171

Technical requirements	172	Cost optimization recommendations	
FinOps challenges for		for AKS clusters	184
containerized workloads	172	Manage underutilized nodes	184
ACI cost allocation	173	Resizing local disks	186
Introducing Kubecost	174	Reserved instances	186
AKS cost allocation	175	Right-sizing your container requests	187
Showback and chargeback shared		Remedying abandoned workloads	188
AKS clusters	180	Right-size persistent volumes	189
		Summary	190

11

Metric-Driven Cost Optimization 191

Technical requirements	192	MDCO and reservation reporting	
Core principles of MDCO	192	using Power BI	193

Setting thresholds for purchasing
reservations   194

Automated reservation purchases
based on MDCO triggers   202

Summary   206

12

Developing Metrics for Unit Economics   207

Technical requirements   208
What is cloud unit economics?   208
Benefits of cloud unit economics   208

Indirect versus direct cost metrics   209

Tracking costs back to business
benefits   209

Developing metrics for unit
economics   210

Cost per patient encounter   210
Revenue per patient encounter   211
Gross margin per patient encounter   212
Cost per claim   213
Revenue per claim   214

Activity-based cost model   215
Summary   216

13

Case Study – Implementing Metric-Driven Cost Optimization and Unit Economics   217

Case study – Peopledrift Inc.,
a healthcare company   217
Challenges   218
Objectives   218

Solution   218
Benefits   220

Summary   221

Index   223

Other Books You May Enjoy   232

Preface

Hi there! **Financial Operations** (**FinOps**) is an emerging discipline that blends financial management, cloud expertise, and data analytics to optimize cloud spending while still ensuring that organizations have the resources they need to support their business objectives.

Practicing FinOps for Microsoft Azure helps with the following:

- Providing visibility of cloud utilization

- Presenting an opportunity to optimize cloud utilization

- Providing a framework to practice with small changes over time

- Reinvesting money saved to fuel innovation

In today's fast-paced and ever-changing cloud environment, mastering FinOps for Microsoft Azure is becoming a necessity for organizations of all sizes to stay competitive and agile.

There are many resources that cover the FinOps practice; the most popular one is the **FinOps Foundation** (`https://finops.org`). This book aims to cover all the best practices for applying FinOps for the Microsoft Azure cloud using native capabilities, except for container cost allocation.

With my extensive experience in conducting assessments for large customers with significant cloud spending, and utilizing the cloud service providers' assessment questionnaire, I can provide relevant information and guidance on FinOps for Azure's cost savings and business value realization activities.

According to a report by Flexera, 63% of enterprises are now using a dedicated FinOps team to manage their cloud spending, up from 45% in 2020. As FinOps adoption continues, the demand for expert resources also grows.

Possessing an understanding of both the financial and technical aspects of cloud services can be a highly valuable skill in bridging the gap between finance, procurement, engineering, and product management teams.

I wish you all the best to start the journey, and before you know it, you will be an expert.

Who this book is for

Individuals and teams who are responsible for managing the financial aspects of their organization's Azure cloud environment.

The target audience who can benefit from this content is as follows:

- **Finance professionals**: You will learn how to manage cloud spending and optimize costs in Azure

- **Cloud architects**: You will be able to design and implement cost-effective Azure solutions for your organization

- **Software engineers**: You will learn how to develop cost-efficient applications and services in Azure

- **DevOps professionals**: You will gain insights to manage cloud resources and control costs as part of your **continuous integration/continuous deployment (CI/CD)** process

- **IT managers**: You will gain visibility into your organization's cloud spending and make informed decisions about resource allocation

- **Business leaders**: You will gain an understanding of the financial implications of your organization's cloud strategy and make strategic decisions based on cost data

What this book covers

Chapter 1, Bringing Visibility and Allocating Cost, discusses the challenges faced by FinOps teams in establishing successful FinOps practices and provides solutions to overcome them. It also highlights Microsoft's Cost Management + Billing tool to gain visibility of the existing IT environment and current cloud cost. The chapter also covers the topics of cost allocation using accounts, management groups, subscriptions, and tags, and explores cost analysis in the Azure portal for FinOps.

Chapter 2, Benchmarking Current Spend and Establishing Budget, provides an overview of benchmarking cloud spend and developing KPIs for consistent reporting to improve performance. It covers topics such as Azure's on-demand and elastic nature, creating budgets, and configuring anomaly alerts in the Azure Cost Analysis tool.

Chapter 3, Forecasting the Future Spend, explains the importance of financial forecasting for FinOps teams and provides insights into various ways to obtain past and current usage and charges for cloud services in Azure. It highlights the significance of tagging resources and covers topics such as setting up cost management connectors in Power BI, forecasting based on manual estimates and past usage, advanced forecasting by application, and fully loaded cost forecasting.

Chapter 4, Case Study – Beginning the Azure FinOps Journey, is a case study of Peopledrift Healthcare's journey to implement FinOps practice and achieve its objectives. The company faced challenges in estimating current and future spending and accurately forecasting spend. By implementing FinOps, they were able to overcome these challenges and improve their financial management gradually.

Chapter 5, Hitting the Goals for Usage Optimization, focuses on the *Usage Optimization* aspect of the FinOps *Optimize* phase, which targets cost avoidance and right-sizing. Cost avoidance can be achieved by deleting unneeded resources, while right-sizing involves selecting the right service SKUs for optimal workload performance. The chapter covers the Project Management Triangle Method for goal setting, as well as setting **objectives and key results (OKRs)** and KPIs. The top 10 usage optimization targets are discussed, along with trade-offs between cost, security, performance, and reliability.

Chapter 6, Rate Optimization with Discounts and Reservations, discusses rate optimization in FinOps, which involves getting better enterprise discounts and purchasing reservations to save costs. The chapter explains enterprise agreements, Azure Advisor recommendations, identifying opportunities for reservations, and monitoring reservation utilization. It also covers reservation purchase and cadence, details, renewal, savings, and chargeback reports, as well as reservation exchange and cancellation.

Chapter 7, Leveraging Optimization Strategies, discusses the importance of utilizing various optimization strategies in a holistic manner, including removing waste, right-sizing, purchasing reservations, savings plans, and highly discounted spot VMs. The focus is on highly discounted spot VMs. Additionally, the chapter discusses the Spot Priority Mix and Savings Plans, which provide a consistent compute capacity with additional spot VMs and are an alternative to reservations, respectively.

Chapter 8, Case Study – Realize Savings and Apply Optimizations, showcases how Peopledrift Inc. adopted the Microsoft Azure cloud and migrated their workload from on-premises to the cloud. After consistently high usage, the FinOps team started looking for ways to save money through rate and usage optimization. The case study discusses the KPIs designed to measure progress, the execution of the usage and rate optimization programs, and the consideration of Azure Savings Plans.

Chapter 9, Building a FinOps Culture, provides an overview of building a culture of FinOps through collaboration across business boundaries. Management buy-in is essential to establish a **Center of Excellence (CoE)** for cloud cost management, which is responsible for bringing stakeholders together and preparing a business plan that articulates savings opportunities.

Chapter 10, Allocating Cost for Containers, focuses on allocating cost for container workloads, specifically in **Azure Kubernetes Services (AKS)** clusters. The challenges of allocating costs in microservices and shared AKS clusters are discussed. The open source tool Kubecost is explored as an industry standard for cost allocation and visibility. The chapter also covers showback and chargeback mechanisms using Kubecost and provides cost optimization recommendations for AKS clusters.

Chapter 11, Metric-Driven Cost Optimization, introduces a cost management strategy for cloud computing environments, which involves using data and analytics to continuously monitor, measure, and optimize cloud costs. The chapter covers the core principles of **Metric-Driven Cost Optimization (MDCO)**, reservation reporting using Power BI, setting thresholds for purchasing reservations, and automated reservation purchases based on MDCO triggers.

Chapter 12, Developing Metrics for Unit Economics, discusses the concept of unit economics in cloud FinOps and its importance in analyzing the costs and revenue associated with delivering a single unit or product within a business. It explains how FinOps teams can use this analysis to make informed decisions about pricing, cost optimization, and resource allocation. The chapter also covers tracking costs back to business benefits, developing metrics for unit economics, and implementing an activity-based cost model.

Chapter 13, Case Study – Implementing Metric-Driven Cost Optimizations and Unit Economics, focuses on implementing MDCO and unit economics for container allocation in the AKS platform. The team utilized a data-driven decision-making process to optimize cloud usage and purchase reservations. They also calculated IT costs per unit of service to understand the profitability of the DeliverNow platform. The case study provides insights into cost allocation for containers and shared services, metrics for reservation purchases, and unit metrics for calculating the per-unit cost and profitability of the business.

To get the most out of this book

You will need to have a basic understanding of Azure cloud services and architecture as well as cloud cost management principles, financial management concepts, and the Azure Cost Management tool. Having access to usage and billing data for your Azure resources is also helpful to analyze your cloud cost.

Software/hardware covered in the book	Operating system requirements
Microsoft Azure portal access	Windows, macOS, or Linux
Microsoft Power BI Desktop	
DAX Studio	
Microsoft Power Automate	
Azure SQL	
AKS	
Kubecost	

You will need a Microsoft Power BI Pro or Premium license to run the data-driven alert feature described in *Chapter 11, Metric-Driven Cost Optimization*.

If you are using the digital version of this book, we advise you to type the code yourself or access the code from the book's GitHub repository (a link is available in the next section). Doing so will help you avoid any potential errors related to the copying and pasting of code.

Download the color images

We also provide a PDF file that has color images of the screenshots and diagrams used in this book. You can download it here: `https://packt.link/YI44t`.

Conventions used

There are a number of text conventions used throughout this book.

`Code in text`: Indicates code words in text, database table names, folder names, filenames, file extensions, pathnames, dummy URLs, user input, and Twitter handles. Here is an example: "Use `Workbook 2`, which provides a list of right-size recommendations for your virtual machines and virtual machine scale sets."

A block of code is set as follows:

```
ResourceContainers| where type =~ 'Microsoft.Resources/
subscriptions'
| extend SubscriptionName=name
| join  ( ResourceContainers | where type =~ 'microsoft.
resources/subscriptions/resourcegroups' | where tags =~ '' or
tags =~ '{}'
| extend resourceGroupName=id, RGLocation=location) on
subscriptionId
| project resourceGroupName, RGLocation, SubscriptionName
```

Any command-line input or output is written as follows:

```
Set-AzVM -ResourceGroupName "mySpotRG" -Name "mySpotVM"
-SimulateEviction
```

Bold: Indicates a new term, an important word, or words that you see onscreen. For instance, words in menus or dialog boxes appear in **bold**. Here is an example: "On the **Basics** tab, select **Flexible** for the **Orchestration** mode."

> **Tips or important notes**
> Appear like this.

Get in touch

Feedback from our readers is always welcome.

General feedback: If you have questions about any aspect of this book, email us at `customercare@packtpub.com` and mention the book title in the subject of your message.

Errata: Although we have taken every care to ensure the accuracy of our content, mistakes do happen. If you have found a mistake in this book, we would be grateful if you would report this to us. Please visit `www.packtpub.com/support/errata` and fill in the form.

Piracy: If you come across any illegal copies of our works in any form on the internet, we would be grateful if you would provide us with the location address or website name. Please contact us at `copyright@packt.com` with a link to the material.

If you are interested in becoming an author: If there is a topic that you have expertise in and you are interested in either writing or contributing to a book, please visit `authors.packtpub.com`.

Share Your Thoughts

Once you've read *FinOps Handbook for Microsoft Azure*, we'd love to hear your thoughts! Scan the QR code below to go straight to the Amazon review page for this book and share your feedback.

`https://packt.link/r/1-801-81016-8`

Your review is important to us and the tech community and will help us make sure we're delivering excellent quality content.

Download a free PDF copy of this book

Thanks for purchasing this book!

Do you like to read on the go but are unable to carry your print books everywhere? Is your eBook purchase not compatible with the device of your choice?

Don't worry, now with every Packt book you get a DRM-free PDF version of that book at no cost.

Read anywhere, any place, on any device. Search, copy, and paste code from your favorite technical books directly into your application.

The perks don't stop there, you can get exclusive access to discounts, newsletters, and great free content in your inbox daily

Follow these simple steps to get the benefits:

1. Scan the QR code or visit the link below

https://packt.link/free-ebook/9781801810166

2. Submit your proof of purchase

3. That's it! We'll send your free PDF and other benefits to your email directly

Part 1:
Inform

This part provides an introduction to FinOps, covering strategies for improving visibility into cloud workloads and techniques for effective cost allocation. Additionally, you will discover how to establish a baseline spend and create and track a budget against spending, as well as techniques for forecasting future spending in the cloud. To solidify your understanding, a case study is presented at the end.

This part contains the following chapters:

- *Chapter 1, Bringing Visibility and Allocating Cost*
- *Chapter 2, Benchmarking Current Spend and Establishing Budget*
- *Chapter 3, Forecasting the Future Spend*
- *Chapter 4, Case Study – Beginning the Azure FinOps Journey*

1

Bringing Visibility and Allocating Cost

With so many things to look at, often, the FinOps team does not know where to start, what to look for, and how to establish a successful FinOps practice for the organization. In this chapter, we will start by bringing visibility to your existing IT environment using two powerful tools. First is the Microsoft **Well-Architected Framework (WAF)** Cost Optimization assessment questionnaire. This will help you understand the people side of things and what practices engineers, DevOps, product owners, and solution architects follow to build and deploy their workloads in Azure. At the end of the self-assessment, you will be provided with a baseline score. This score is your starting point to keep track of your improvements over a quarter or year. As you make small adjustments, this score will improve over time. The second tool we will explore is Microsoft's Cost Management + Billing to show your current cloud costs and the drivers behind them.

Once you have this visibility, we will move on to understanding cost allocation. We will use accounts, management groups, subscriptions, and tags to allocate the cost proposed by the finance team.

In this chapter, we're going to cover the following main topics:

- Tools used in this book for implementing FinOps for Microsoft Azure
- What is the Microsoft Azure Well-Architected Framework?
- Creating a baseline using WAF – the Cost Optimization assessment
- Cost allocation from an accounting point of view
- Cost allocation in Azure for FinOps
- Exploring cost analysis in the Azure portal

Let's get started!

Technical requirements

We will be using the Azure Well-Architected Review from Microsoft Assessments to accomplish the tasks in this chapter, which is available at `https://docs.microsoft.com/en-us/assessments`. It is an online self-assessment that takes about 30-60 minutes to complete.

When using this tool, be sure to sign in using your organization ID. In the next section, we will look at the tools that need to be installed to complete the tasks throughout this book.

Tools used in this book for implementing FinOps for Microsoft Azure

Your organization uses Microsoft Azure. You will be using Microsoft Azure's native capabilities to fulfill the requests by the FinOps team. Please follow the instructions provided to set up the tools and verify your access in Microsoft's Azure portal.

Azure CLI

We will be using Microsoft Azure's command-line interface to run ad hoc scripts. Follow these steps to download and install the `az` command-line interface:

1. Open the Microsoft Edge browser.
2. Navigate to `https://aka.ms/installazurecliwindows`.
3. Once the download is complete, navigate to the `Downloads` folder.
4. Run the MSI installer.
5. Once complete, open Command Prompt and run the following command:

    ```
    az login
    ```

This command will open a pop-up window where you can sign in to your Azure subscription. You will see the subscription details in Command Prompt after successful authentication.

Power BI Desktop

Power BI Desktop is Microsoft's tool for providing rich interactive reports and visual analytics. This tool is free to use and provides built-in connectors for Azure Cost Analysis. Follow these steps to download and install Power BI Desktop:

1. Open the Microsoft Edge browser.
2. Navigate to `https://aka.ms/pbidesktopstore`. This link will open the Windows Store.

3. Click **Install**.

4. Once the installation is complete, go to the Start menu and open **Power BI Desktop**.

You will see Power BI Desktop's welcome screen, which contains *Getting started* videos and links to the Power BI tutorials.

Azure Cost Management + Billing

Cost Management + Billing is a Microsoft native tool built into the Azure portal that allows you to analyze, manage, and optimize the costs of your application or workload. It currently supports the Microsoft Online Services Program, Enterprise Agreement, and Microsoft Customer Agreement types of billing account. To access Cost Management + Billing, you will need the **Reader** or **Cost Management Reader** permission in Azure. Let's verify that you can access the billing data:

1. Open the Microsoft Edge browser.

2. Navigate to `https://portal.azure.com` and sign in with your organization's account.

3. In the top search bar, search for `Cost Management + Billing` and select the highlighted service.

4. Once it opens, you should be able to see the overview page, where you can see the latest billed amount, invoices over time, and subscriptions.

Cost Management + Billing has a tab on the left where you can access a more detailed cost management view. You can also give it a try and make sure you can access the page.

Azure Advisor

Azure Advisor is a personalized service that scans Azure resources proactively periodically and provides actionable recommendations for workload optimization. To access the Advisor recommendations, you will need at least the **Reader** permission in Azure.

Follow these steps to access Azure Advisor:

1. Open the Microsoft Edge browser.

2. Navigate to `https://portal.azure.com` and sign in with your organization's account.

3. In the top search bar, search for `Advisor` and select the highlighted service.

4. Once it opens, you should be able to see the overview page with your Advisor score, score history, and recommendations across five categories – **Cost**, **Security**, **Reliability**, **Operational Excellence**, and **Performance**.

You can click **All Recommendations** on the left navigation bar to view recommendations across all categories.

Azure Monitor

Azure Monitor is a cloud-based monitoring and analytics service by Microsoft. It collects and analyzes telemetry data from various sources, such as applications, infrastructure, and network devices. It helps you identify and troubleshoot issues proactively, optimize performance, and gain insights for better decision-making. At a minimum, you will need the **Reader** permission in Azure to view the monitoring data and dashboards.

Follow these steps to access Azure Monitor:

1. Open the Microsoft Edge browser.
2. Navigate to `https://portal.azure.com` and sign in with your organization's account.
3. In the top search bar, search for `Monitor` and select the highlighted service.
4. Once it has opened, you will be presented with an overview page that displays your monitor's **Insights**, **Detection**, **Triage**, and **Diagnostics** details.

On the overview page, you can also use the curated monitoring views for specific Azure resources, such as **Application Insights**, **Container Insights**, and **VM Insights**.

Azure Pricing Calculator

Azure Pricing Calculator is an online tool that you can use to estimate the cost of your workload. Once you know what services you will need, you can select those in Pricing Calculator; you can select **SKUs** and **Instance Configuration** to get the monthly price. Here, you can also get estimates for pay-as-you-go, 1-year reservation, and 3-year reservation pricing, which includes a 30-60% discount depending on the service.

To access Azure Pricing Calculator, do the following:

1. Open the **Microsoft Edge** browser.
2. Navigate to `https://azure.microsoft.com/calculator` and log in to Pricing Calculator with your organization's account.
3. Once you've logged in, you will be able to create, save, and share the pricing estimates.

Note that, by default, Pricing Calculator shows the street price. Your organization may have a special licensing agreement with Microsoft, for example, Enterprise Agreement. To get your discounted Enterprise Agreement pricing, select your licensing program under the **Support** section of the page. The pricing will be updated once you select the correct agreement. Next, let's move on to the Microsoft WAF.

What is the Microsoft Azure Well-Architected Framework?

The Azure WAF is a guiding principle across five pillars of architectural excellence that produces high-quality and efficient architecture for your workload in Azure. These pillars are as follows:

- Reliability

- Security

- Cost Optimization

- Operational Excellence

- Performance Efficiency

Since the FinOps team is focused on managing the cloud costs and maximizing the value delivered to the business, we will focus on the Cost Optimization pillar of WAF. WAF is supported by six elements – Well-Architected Review, Azure Advisor, Documentation, Partner Support and Service Offer, Reference Architectures, and Design Principles. Azure Well-Architected Review examines your workload through the lens of cost management. WAF reviews also help bring visibility to engineering, the product and DevOps teams' knowledge, and the ability and priorities around considering the cost for every decision they make for the cloud infrastructure. Let's assemble the team and have them take the WAF Cost Optimization assessment.

Creating a baseline using the WAF Cost Optimization assessment

The scope of the WAF Cost Optimization assessment is per application or workload. First, you must identify the large workload. This assessment is a collaborative effort between FinOps, cloud governance, engineering, DevOps, architecture, and product teams. Ideally, this assessment is completed in a single two-hour meeting. The FinOps lead facilitates the meeting, provides clarification of the questions, and records the final answers. As you go through the questions, you will realize how much knowledge flows through. This is a fantastic opportunity for you to get a good insight into the thinking and decision-making patterns of engineering and DevOps teams. Follow these steps to run through the assessment:

1. Open the Microsoft Edge browser.

2. Navigate to `https://docs.microsoft.com/en-us/assessments`.

3.   Select **Azure Well-Architected Review** from the **Available assessments** list:

Figure 1.1 – Microsoft Assessments

4.   To save your progress, you will need to *sign in* with your organization ID. After that, give your *assessment* a name and click **Import** to include Azure Advisor recommendations. When you click **Import**, the pop-up window will list all your Azure accounts. Select the account that belongs to the workload and then click **Continue**. On the next screen, select the workload subscription and click **Import**.

Figure 1.2 – Azure Well-Architected Review

5. Once your subscription has been linked, you will see a screen similar to the following. At this point, click **Start**.

Figure 1.3 – Import Azure Advisor recommendations

6. The Cost Optimization pillar falls under **Core Well-Architected Review**. So, check that option and then click **Next**.

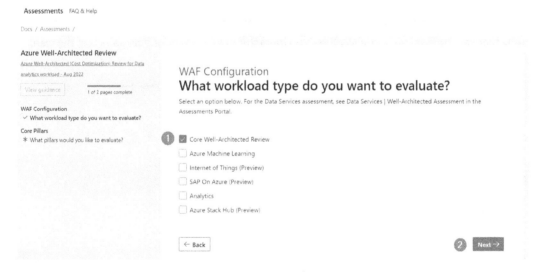

Figure 1.4 – WAF Configuration

7.   On the next screen, select **Cost** and then click **Next**.

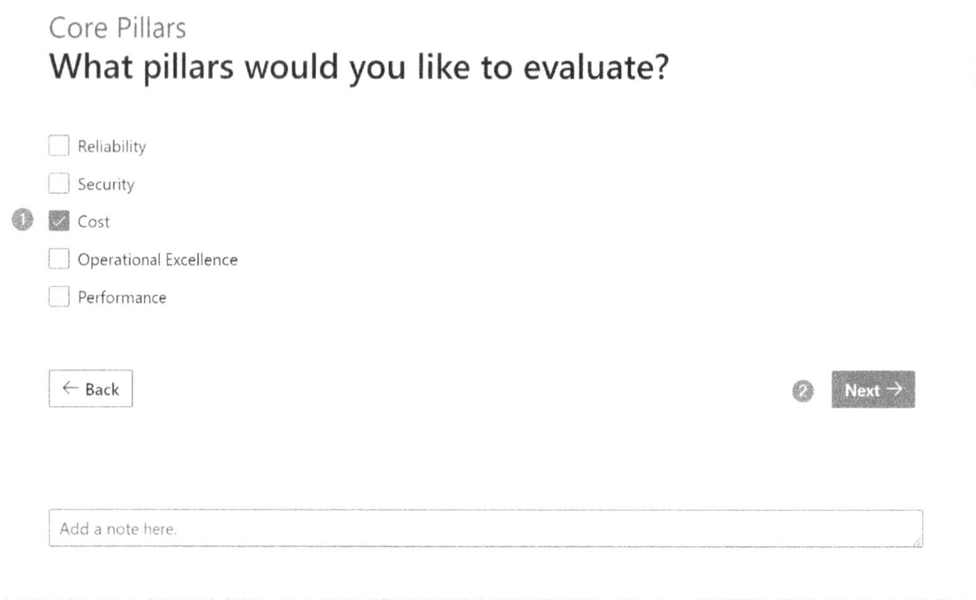

Core Pillars
What pillars would you like to evaluate?

☐ Reliability

☐ Security

① ☑ Cost

☐ Operational Excellence

☐ Performance

← Back   ② Next →

Add a note here.

Figure 1.5 – Core pillars of WAF

At this point, the WAF cost assessment questions start. Let's understand the key parts of the assessment screen. There are nine sections, and each section has a different number of questions that the team will collaboratively answer.

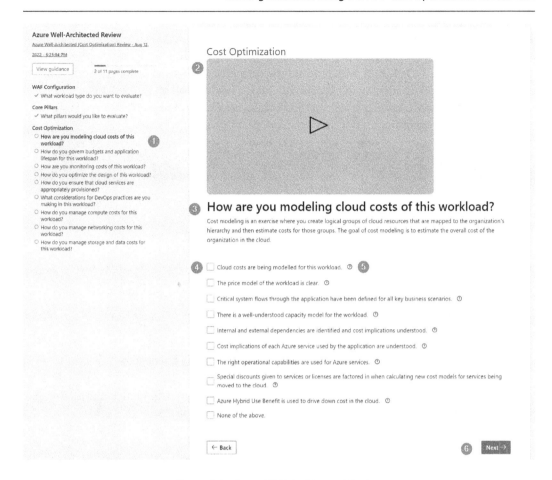

Figure 1.6 – Azure Well-Architected Review

The main parts of the assessment screen are as follows:

1. On the left-hand side is the list of **Cost Optimization** sections.

2. At the top of the screen is a short video explaining what each section means.

3. Under the video is the title of the current section – for example, **How are you modeling cloud costs of this workload?**.

4. Below is the list of questions, each with a checkbox. You discuss these questions with the team and check the box if applicable. If they are unsure, leave the box unchecked.

5. The question marks to the right of each item provide more context to help you understand the option.

6. Once you have completed all the questions, click **Next** to move to the next section.

Once you have completed all the sections, the **Your results** screen will appear. Let's understand the results screen. The top part of the screen displays your overall score. As shown in the following screenshot, I received a score of **MODERATE**, which means a score between 33 and 76. So, my workload is following some cost optimization best practices, but there is lots of room for improvement. You can export the result to a CSV file and use that later to either create your own dashboard to keep track of the progress on each item or import it into Azure DevOps for tracking purposes.

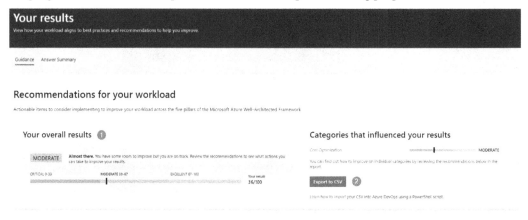

Figure 1.7 – Overall results of the Well-Architected Review assessment

Below the results section, the screen shows a list of all recommendations with a link to the Microsoft documentation, which helps you practice that specific cost optimization area.

Figure 1.8 – Detailed results of the Well-Architected Review assessment

With that, we have looked at how to create a WAF baseline score. Now that you have the baseline score for your selected workload, let's understand what cost allocation is and how you can practice it in Azure.

Cost allocation from an accounting point of view

As per the Oxford dictionary definition, cost accounting involves recording all the costs incurred in a business in a way that can be used to improve its management. There are two approaches to cost allocation in cost accounting: traditional and **activity-based cost (ABC)**.

In traditional cost accounting, we have a plant-wide overhead rate. It is calculated using machine hours or labor hours. This will give us one rate that we can allocate throughout the year. Due to this simplicity, traditional costing is easier to implement.

In traditional costing, only the product cost is allocated. It only considers manufacturing overhead. **Selling, general, and administrative (SG&A)** costs are not considered in traditional costing. Traditional costing can be used to calculate the cost of goods sold and it is an accepted method by auditors. But there are disadvantages to the traditional costing method – for example, it only factors in one rate. Let's take a scenario where one product is causing issues and requires a lot more customer support hours. In that case, we are understating our cost because we did not allocate the cost to that product line.

Cost accounting is all about making quality decisions, and ABC allocation is the method that will provide us with accurate costs. ABC will compute the rate for each cost pool as that activity occurs. Once we have the accurate cost, we can make the best decision on whether to keep the product line, drop it, or optimize certain processes.

Having covered the advantages of ABC, let's now look at it in detail.

What is ABC allocation?

ABC in cost accounting is a method of accurately spreading the cost to the relevant business activities so that overhead costs can be calculated in proportion to the rate of the activities.

Let's understand the key definitions first before we look at an example of ABC allocation:

- **Product**: This is an item or a service that a business sells to its customers. These products can be physical or virtual. Backpacks, suitcases, furniture, and computers are examples of physical products, while email, credit card processing, and appointment services are examples of virtual products.

- **Overhead**: This is an expense that a business must pay so that the business can continue operating. Rent, electricity, admin, and customer support are examples of overhead expenses.

- **Cost pool**: These are business process activities that are grouped into a pool for cost allocation. Each organization defines the most relevant cost pools to their business needs. Assembly, order processing, marketing, and customer support are examples of cost pools.

- **Measure**: This is a driver for the cost pool. The most common example of a measure is the number of units, number of customers, or number of orders.

- **Activity rate**: This is calculated by dividing *cost* by *activity*. It is the per-unit cost of the activity – for example, the assembly cost of 500,000/20,000 units = $25 per unit activity rate.

To better understand this, here is an example of *World Travel Inc.*, which manufactures travel backpacks and suitcases. World Travel Inc. has an overhead expense of $1,000,000 that needs to be allocated to these product lines. The line manager has identified the following cost pools, measures, and weights:

Cost Pool	Measure	Weight
Assembly	# of Units	50%
Order Processing	# of Orders	10%
Customer Support	# of Customers	25%
Other	N/A	15%

Now, we can calculate the ABC for the $1,000,000 overhead cost as follows:

Cost Pool	Cost	Measure	Activity Rate
Assembly	$500,000	20,000 Units	$25 per Unit
Order Processing	$100,000	2,000 Orders	$50 per Order
Customer Support	$250,000	5,000 Customers	$50 per Customer
Other	$150,000	N/A	N/A

We can use the activity rate to calculate the cost of backpacks and suitcases:

	Backpacks		Suitcases	
Assembly	10,000 Units	$250,000	10,000 Units	$250,000
Order Processing	700 Orders	$35,000	300 Orders	$15,000
Customer Support	150 Customers	$7,500	50 Customers	$2,500

As you can see in the preceding calculation, for 10,000 units of orders for two different product lines, the order processing and customer support costs are different. This is the exact problem that ABC solves. In conclusion, the main driver for implementing ABC is that it is much more accurate than traditional costing, which takes one plant-wide overhead rate and asserts that overhead is just driven by machine hours or labor hours. ABC is the method that will give an accurate cost for the product. When we have the most accurate cost, we can make the best decision about whether to keep the product line or optimize the operations.

How do we allocate costs in Azure? We'll explore that in the next section.

Cost allocation in Azure for FinOps

The goal of the FinOps team is to create visibility into cloud spending and enable granular cost allocation to create shared accountability of cloud spending. Proper cost allocation will help teams to see their cloud spending and the impact their action or inaction has on the bill. Spending data must be properly mapped to the organizational hierarchy by cost center, applications, and business units by using the account hierarchy and resource tagging.

Cost allocation using the account, management group, and subscriptions hierarchy

The billing account, department, and account hierarchy for an Enterprise Agreement offer type allows you to organize and allocate your costs at the highest level. In the following example, the organization has created two departments (**Marketing** and **HR**) under **Billing Account**. Invoices are generated at the **Billing Account** level. The **Department** part groups the accounts (that contain subscriptions) to organize costs into a logical grouping. The EA admin can configure the spending quota for each department. This is an excellent place to implement a budget. The enterprise administrator and the department administrator will receive notifications once the quota has reached 50%, 75%, 90%, and 100%.

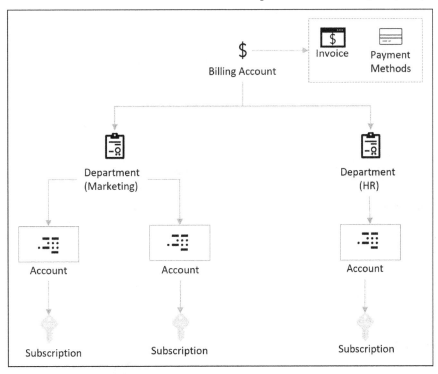

Figure 1.9 – Azure Enterprise Agreement's department hierarchy

While you can use Enterprise Agreement's Department hierarchy for cost allocation, in practice, it does not provide the granularity that the FinOps team is looking for.

> **Note**
>
> To learn more about the **Microsoft Customer Agreement** (**MCA**) billing hierarchy, please refer to the Microsoft documentation at `https://docs.microsoft.com/en-us/azure/ cost-management-billing/manage/view-all-accounts`.

Here, management groups come into the picture. As you can see in *Figure 1.10*, we are starting from the top and moving down toward a bit more granularity. Management groups allow efficient access, policy, and compliance for hundreds of subscriptions. While they are not directly related to helping with cost allocation, the latest cost analysis scope object now includes filtering by using management groups. So, now, you can group the cost of all resources under a single management group and report it back to the business. The following is an example of a management group hierarchy that helps with cost allocation:

Figure 1.10 – Organizing subscriptions under management groups

In the preceding example, the first level of management group is called the **Root** management group. Management groups can be six levels deep. Under **Root**, we have **Organization**, and under **Organization**, we have **Platform**, **Landing Zone**, **Decommissioned**, and **Sandbox** management groups. Under the **Platform** management group, we have **Identity**, **Management**, and **Connectivity**. Each has a corresponding **Subscriptions** group.

Using this management group structure, the FinOps team can easily group the cost by a given management group. For example, to know how much the SAP workload costs, you can go to the **Cost Analysis** dashboard and select the scope of the SAP management group. That will include all the subscriptions under SAP and provide the total cost.

Subscriptions in Azure support tags. To allocate cost at the subscription level, you can assign tags to the subscription and group the cost by the given tag. For example, by using tags on subscriptions, you can easily allocate and get the total cost incurred by a given tag on a subscription.

Cost allocation using resources tags

Tags are essentially metadata that you can apply to Azure resources in the form of key/value pairs. For example, if you want to distinguish the cost of the development environment from the production environment for the marketing application, you can assign the `Environment = Development` or `Environment = Production` tag to all the resources in Azure.

It is important to note that each resource, resource group, and subscription can have a maximum of 50 tags. If you need more than 50 tags, you can use a JSON string in the tag value. Also, note that Azure Automation, Azure **Content Delivery Network** (**CDN**), and Azure DNS (zone and A records) only support a maximum of 15 tags.

> **Note**
>
> Not all resource types in Azure support tags; these resources are called *untaggable*. To find out whether a resource supports tags or not, refer to the Microsoft documentation at `https://docs.microsoft.com/en-us/azure/azure-resource-manager/management/tag-support`. Please see the case study for the strategy to allocate costs for untagged or untaggable resources.

Depending on the FinOps team's maturity level, the following are the minimum tags that should be applied to the resources. To enforce tag coverage, you can create an Azure policy to audit or deny the resource creation if the required tags are not provided:

Tag Name	Description
`appid`	`appid` can be any unique ID that an organization can use to distinguish applications. The most common values are application IDs from CMDB, as in the following example: `appid = 141788`
`env`	This represents the application environment, as in the following examples: `env = dev` `env = prod`
`department`	This represents the department that owns the application for billing purposes, as in the following example: `department = marketing`
`itowner`	This represents the IT owner of the application, as in the following example: `itowner=james@corp.com`

Tag Name	Description
`businessowner`	This represents the business owner of the application, as in the following example: `businessowner=mike@corp.com`
`costcenter`	This represents the organization's cost center code for this application, as in the following example: `costcenter=10101`

There are various ways you can apply these tags to Azure resources. The most common and effective way is to embed tags in your **infrastructure as code** (**IaC**) solution (Bicep or Terraform). Alternatively, you can use the Azure portal or Azure CLI commands to list, add, update, or delete tags.

To add tags to a Terraform template for a resource group, use the following code:

```
resource "azurerm_resource_group" "rg-dev" {
  name     = "rg-marketing-website-dev"
  location = "eastus"
  tags = {
    appid = "141788"
    env = "dev"
    department = "marketing"
    costcenter = "10101"
    itowner = "james@corp.com"
    businessowner = "charles@corp.com"
  }
}
```

To manage tags with PowerShell and Azure CLI commands, please check out the Microsoft documentation at `https://docs.microsoft.com/en-us/azure/azure-resource-manager/management/tag-resources`.

In the next section, we will look at how to view the allocated cost by tag using Azure's cost analysis tool.

Exploring cost analysis in the Azure portal

Microsoft Azure has the most advanced, free, in-portal capability to analyze the cost of your workload. Let's look at the Cost Management + Billing feature in the portal. You will learn how to do the following:

- Identify the offer type for your subscription(s)
- View the accumulated and forecasted cost

- View the cost grouped by service

- View the cost grouped by management group

- View the cost grouped by tag

- View the create, save, and share custom cost analysis views

Let's look at how to accomplish these tasks using the Azure Cost Management + Billing feature in the portal.

Identifying the offer type for your subscription(s)

When you purchase or create a Microsoft Azure subscription, an **Azure offer** governs the payment terms and conditions, service rate, exclusions, payment options, cancellation policy, and service agreement between your organization and Microsoft. Here is how you can find out the *offer type* in the Azure portal:

1. Open the Microsoft Edge browser.

2. Navigate to `https://portal.azure.com/#view/Microsoft_Azure_ CostManagement/Menu/~/overview` and then click on **Cost Analysis**.

3. Click on **Properties**, under the **Settings** section. The **Type** field indicates your subscription's offer type. The following screenshot shows that the subscription has an *MCA offer type*.

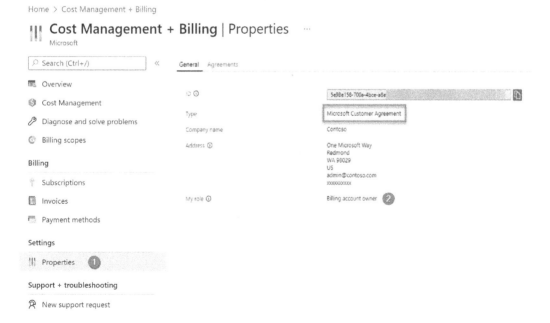

Figure 1.11 – Azure Cost Management + Billing

There are many different types of offers available in Azure. To view the full list and details about each offer, please check out `https://azure.microsoft.com/en-us/support/legal/offer-details/`.

Accumulated and forecasted cost

One of the most common requests from the finance and procurement team is to provide accumulated and forecasted costs for the workloads running in Azure. To accomplish this request, you can use the built-in accumulated cost view. Furthermore, you can share the view directly with the finance and procurement team or export the data in PNG, Excel, or CSV format for further analysis. The forecasted cost is based on your historical resource usage and shows a prediction of your estimated cost for up to a year.

Open **Cost Management + Billing** in the Azure portal and then click on **Cost Analysis**. The default **Cost Analysis** view will display actual and forecasted costs.

Figure 1.12 – The Azure Cost Analysis feature

Since this is the first time we are looking at Cost Management + Billing, let's get familiar with its key features:

1. You can create many different views with customizations such as date ranges, scope, filter by service or tag, group by location or resource, and so on. Then, using the **Save** button at the top, you can create a new customized view.

2. To share the view directly, use the **Share** button. It provides a direct URL to the cost management view. Remember that appropriate RBAC permissions will be required for the viewer to access the report.

3. Using the **Download** button, you can download the image or data (as a CSV or Excel file) and share it with others.

4. The **Scope** filter allows you to restrict the cost analysis to a subscription or management group.

5. The **VIEW** dropdown shows built-in views such as accumulated cost, daily cost, cost by service, and any custom views that you may have saved earlier.

6. The *date range* filter provides options to view your costs daily, monthly, quarterly, or yearly, or select a custom date range.

7. The **ACTUAL COST (USD)** option shows the total cost. For Enterprise Agreement accounts, this includes all usage, reservations, and marketplace purchases.

8. The **FORECAST: CHART VIEW ON** option shows the predicted cost based on your usage in the past.

9. The dark color shows your accumulated cost.

10. The light color shows your forecasted cost.

Now, you can provide the actual resource cost by selecting the right scope and date range.

We will take a deeper look at forecasting cost in *Chapter 3, Forecasting the Future Spend*.

Cost grouped by service

In this task, we will drill down the cost to the service level to better understand the parts of your infrastructure that cost more. This view helps you understand the primary cost drivers and enables you to adjust the service usage to control the cost:

1. Open **Cost Management + Billing** in the Azure portal and then click on **Cost Analysis**.

2. Change the view to **CostByService** and select the appropriate date range.

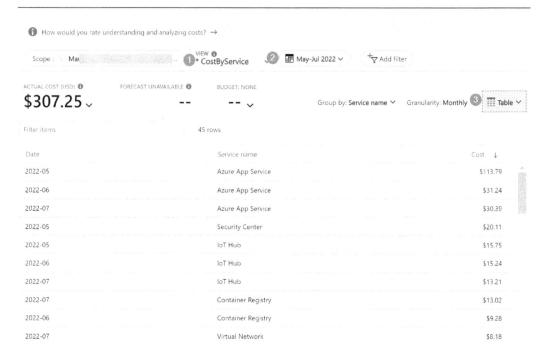

Figure 1.13 – Cost grouped by services

3.   Set the view to **Table** to see the data in table format.

Often, you will want to see the networking (virtual network, bandwidth, and so on) versus compute versus storage cost. This view provides all these details and enables you to create custom reports by exporting data into Excel.

Cost grouped by management group

Management groups in Azure provide you with an efficient way to manage resources, access, policies, and compliance by logically grouping the subscriptions. The management group hierarchy has also extended to Cost Management + Billing. Now, the cost management scope filter supports management group-based filtering. Let's look at an example. The engineering vice president wants to know how much money we are spending to host corporate intranet applications versus customer-facing online business applications to calculate the overhead cost. To provide this data, follow these steps:

1.   Open **Cost Management + Billing** in the Azure portal and then click on **Cost analysis**.

2.   Change the **Scope** filter to -**corp management group** under the landing zone. This will provide costs for all the subscriptions hosting internal corporate applications.

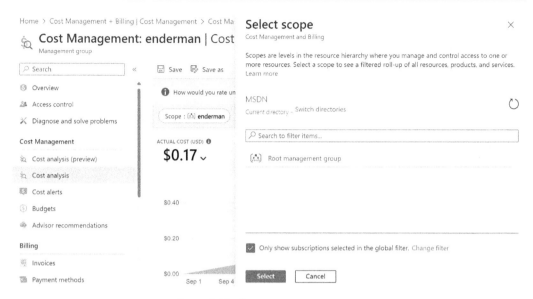

Figure 1.14 – Root management group

3. To provide the cost for online business-facing applications, change the scope to the online management group under the landing zone.

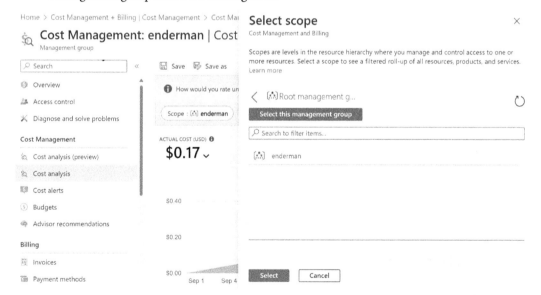

Figure 1.15 – Child management group

Please note that the aforementioned steps assume you have organized your subscriptions into management groups according to the Microsoft WAF guidance. More details are available at `https://docs.microsoft.com/en-us/learn/modules/enterprise-scale-introduction/3-enterprise-scale-architecture-approach`.

Cost grouped by tag

As you have seen previously, in Azure, you can use resource tags to allocate costs. After careful planning, your team has set up appropriate tags for the workload. In this example, the `env:dev` or `env:prod` tag has been applied to each resource in the `dev` and `prod` subscriptions for the marketing website's workload. Now, for the annual budget planning, the finance team is asking you to provide costs for the dev and production environments. Here is how you can accomplish this task:

1. Open **Cost Management + Billing** in the Azure portal and then click on **Cost Analysis**.

2. Click on the **Group by** dropdown, then select **Tag** as a filter. Next, select a given tag to group by the cost. In this case, I have selected the **env** tag, which contains values such as **dev** and **prod**.

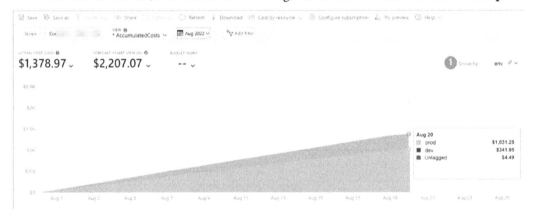

Figure 1.16 – Cost grouped by the env tag

Note that when you group the cost by tag, you will see the **Untagged** category. This is because there are untagged resources in your subscription. Later, we will discuss how to deal with untagged and untaggable resources since, practically, it is not possible to tag 100% of resources all the time.

Creating, saving, and sharing custom cost analysis views

Finally, you want to create custom views and share them with the FinOps team to standardize the cost showback across teams. You can create private views that only you can see, as well as shared views. Shared views are accessible to anyone who has cost management access. To edit or delete a shared view, you will need cost management contributor access.

The finance team is asking for a standardized report that shows quarterly spending by service in the Azure US West 2 region for showback purposes. This is how you can accomplish the task:

1. Open **Cost Management + Billing** in the Azure portal and then click on **Cost Analysis**.

2. Click on the **VIEW** dropdown and select **CostByService**.

3. Change **Time Range** to **This Quarter**.

4. Click on **Add filter**, then select **Location** from the drop-down list. For **Value**, select **US West 2**. For **Chart Type**, select **Column (stacked)**.

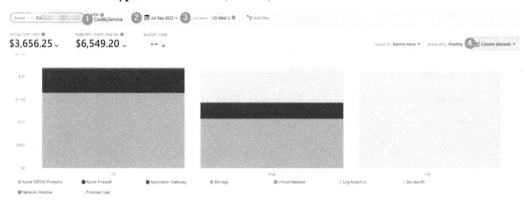

Figure 1.17 – Cost by services custom view

5. To save this view so that team members can access it, click on **Save as** on the top toolbar:

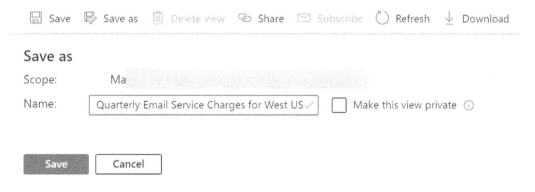

Figure 1.18 – Save the custom dashboard

6. To share the view via a direct URL, click on **Share**.

The shared views will then appear under the **VIEW** dropdown of the **Cost Analysis** page. That concludes this chapter. Let's recap what we've learned so far.

Summary

We have covered lots of things in this chapter. We started by introducing native Azure tools and then looked at the Microsoft WAF Cost Optimization assessment. The team collaboratively works to complete the assessment and establish the baseline score. Then, we discussed traditional cost allocation and ABC allocation from a cost accounting point of view. Taking that concept further, we used billing accounts, management groups, subscriptions, and resource tags to allocate the cost for a workload running in Azure. By using the Cost Analysis feature of Cost Management + Billing in Azure, we looked at various scenarios and how to effectively use the tool to accomplish the tasks given by the finance and procurement teams.

In the next chapter, we will look at benchmarking the current cloud spending and how to establish a budget for a workload, department, or business unit.

2

Benchmarking Current Spend and Establishing Budgets

What can be measured can be improved. In this chapter, we will look at how to establish benchmarking cloud spend within teams, across teams, and across organizations. Benchmarking requires the development of new **Key Performance Indicators** (**KPIs**). This chapter will explain the proven method of successfully developing KPIs. We will look at budget allocation and tracking using the Azure cost analysis tool as well as configuring anomaly alerting for spending higher than the allocated budget.

In this chapter, we are going to cover the following main topics:

- The on-demand and elastic nature of Azure
- Developing KPIs for consistent reporting
- Benchmarking between teams
- Creating and managing budgets in Azure cost analysis
- Creating and managing alerts in Azure cost analysis

Technical requirements

We will be using the following tools to accomplish the tasks in this chapter:

- For Cost Analysis, the Microsoft *Cost Management + Billing* tool is available at `https://portal.azure.com/#view/Microsoft_Azure_CostManagement/Menu/~/overview`. Alternatively, you can also find **Cost Management** + **Billing** by signing in to the Azure Portal and, in the top center search bar, typing `Cost Management + Billing`.

When using these tools, sign in to Azure Portal using your organization ID. Please refer to *Chapter 1* for the minimum RBAC permissions required to use Azure **Cost Management** + **Billing**.

The on-demand and elastic nature of Azure

Azure is designed to scale. Workloads on Azure can easily scale up or down and scale out or in depending on demand. When coming from a data center mindset, where compute, storage, and networking capacity are preallocated and rarely changed, the cloud is a totally different ball game. Azure is all about *pay-as-you-consume*. If you don't consume, you don't have to pay. And this principle can be used to our advantage.

When you think of deploying a workload to Azure, start small. There is no need to pre-provision peak capacity when you can simply auto-scale based on metrics such as CPU, memory, or queue length. Also, consider shutting down development and test environment resources on weekends or holidays based on usage patterns. In this way, you can entirely avoid the cost associated with those resources.

All of this is possible only when you develop sensible KPIs and business metrics. Let's look at how we can achieve that.

Developing KPIs for consistent reporting

A KPI is a quantifiable measure used to evaluate the success of an organization, employee, and more to meet the performance objectives. Put simply, KPIs are outcome-based statements of what you want to achieve and by when. KPIs and business metrics are not the same. KPIs are related to achieving a specific business outcome.

Let's look at two different types of KPIs next.

Strategic and operational KPIs

Strategic KPIs are big-picture indicators to monitor an organization's goal at the highest level. Generally, executives look at one or two strategic KPIs to know how an organization is performing. Subscription revenue, market share, and data center migrations to the cloud are examples of strategic KPIs.

Operational KPIs are indicators of measure for a shorter timeframe and are mainly used for improving organizational processes and efficiencies. Monthly cloud cost and subscription revenue by region are examples of operational KPIs.

Leading and lagging KPIs

Leading KPIs are a measure that predicts future conditions. They are described as input. Leading KPIs define necessary actions to achieve a particular goal with measurable outcomes. Some examples of leading KPIs for a healthcare organization are listed as follows:

- Patient satisfaction score
- Treatment success rate
- Average wait time

A lagging KPI indicates past and current performance measures. The lagging indicator shows the outcome that already has occurred to gain insight for future planning. Some examples of lagging KPIs for a healthcare organization are listed as follows:

- Readmission rate
- Inpatient admissions
- Referral rate

Why do you need KPIs?

In FinOps practice, you might be wondering why we need KPIs for IT. Well, there are three major reasons to define KPIs. First and foremost, properly defined KPIs engage and unify various teams to work toward a common goal. KPIs also communicate long- and short-term strategies to employees to achieve results. Second, KPIs connect a company's culture and purpose. Making money is not a mission and employees will not deeply connect with the company if they don't see the value their work provides. KPIs are a link between your mission and employees and make them feel their work is purposeful and fulfilling. And finally, KPIs make everyone accountable. In FinOps, collaboration without accountability does not bear results. It is important to understand that everyone in IT is accountable for the cost of the cloud.

Defining, measuring, and reporting KPIs

To define a good KPI, it must have five attributes:

- **Measure** – This is the verbal expression. Put simply, it defines what we are measuring. It must be expressive; for example, the number of new patient admissions by end of year.
- **Target** – This is a numerical value that we want to achieve; for example, 2,000 new patients.
- **Source** – This is where the data is coming from. Most organizations have multiple sources and need to identify the accurate source for the data; for example, **Electronic Health Record (EHR) System**.
- **Frequency** – This is how often you will be reporting on this KPI; for example, monthly.
- **Visualization** – This is how to use appropriate bar charts, pie charts, or timelines:

Key performance indicators			
Measure	**# Of new patient admissions by EoY**	**% Patient satisfaction score**	**% Referral Rate**
Target	2,000	90%	80%
Source	EHR	Patient web portal	Patient appointment management system
Frequency	Monthly	Quarterly	Annual
Visualization			

Table 2.1 – KPI example

In *Table 2.1*, we have defined three KPIs:

- The first KPI measure is for getting 2,000 new patients for the current year. The source data for getting the count of new patients will be the EHR system (e.g. EPIC) and the KPI will be updated monthly.

- The second KPI measure is to achieve a 90% patient satisfaction score. The source for this KPI is Patient web portal, which sends out satisfaction survey emails and the KPI will be updated quarterly.

- The third KPI measure is to increase patient referrals by 80%. The data source for this KPI is Patient appointment management system, and the KPI will be updated annually.

KPI reporting and measuring progress toward the goal can be accomplished using Microsoft Power BI. Power BI is an advanced analytics and visualization tool from Microsoft. You could use the desktop version to develop the KPI's dashboard and use Power BI Online to share the report with the rest of the teams.

Now that you have an understanding of example KPIs and their attributes, let's look at benchmarking examples between teams.

Benchmarking between teams

Scorecards are the best tool to benchmark the team's performance to take action against objectives defined in the scorecard. It will allow you to find the teams who are performing and meeting the expectations versus teams that are maintaining the status quo. Depending on your organization's goal, you can collaboratively create scorecards to measure teams. Here are three scorecards as an example:

- Count of instances (VMs) versus containers versus serverless
- Count of resources that have auto-scale policies
- Reserved instances coverage

The first scorecard is *Instances (VMs) versus Containers versus Serverless*. This scorecard tells us which teams are still using IaaS VMs for their workloads. It's a generally accepted practice that you can save money in the cloud by stepping up one level. If you are using IaaS VMs, refactor your application so that it can run on containers. If you are running your apps in containers, explore opportunities to run them serverless (e.g. function apps). This scorecard compares teams based on their count of VM, container, and serverless instances. The team that scores higher on serverless count gets recognition. Teams that are stuck on IaaS workloads and not refactoring and rehosting are asked for justification for not doing enough to move the needle:

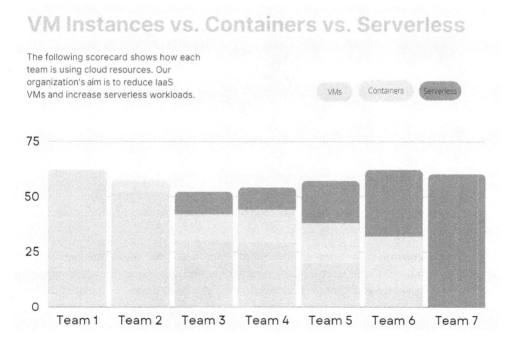

Figure 2.1 – Benchmarking for VM instances versus containers versus serverless

In the preceding example, we can see that **Team 7** has fully adopted a serverless platform and is getting the benefits of dynamic scale. **Team 1** is IaaS- and VM -heavy, and the FinOps team needs to seek more information about their workload and provide alternative cost-saving suggestions such as reservations.

The second scorecard is the count of resources, which has auto-scale policies. Why is this important? Because you want to measure the teams who are taking steps to account for workload utilization patterns and allowing infrastructure to scale in and out based on demand. Dynamic scaling is the true spirit of cloud computing and a very effective tool to avoid cost accumulation when the demand is less, and compute power is over-allocated:

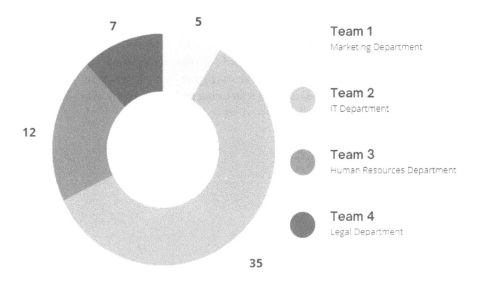

Figure 2.2 – Benchmarking for auto-scale enabled resources

In the preceding example, we can see that **Team 2** has the highest resources with auto-scale enabled, and it is matching the resources with the demand.

The third example scorecard is the reserved instance coverage. Each team is measured against their utilization of reservations. Getting the data for reservations and calculating the coverage can be challenging, but the resource you want to check out first is `https://docs.microsoft.com/en-us/rest/api/billing/enterprise/billing-enterprise-api-reserved-instance-usage`:

Figure 2.3 – Benchmarking for reservation coverage

In the preceding example, we can say that **Team 7** is leveraging 100% reservations for their computer needs, while **Team 1** is not using reservations at all and paying the street prices. The teams in between have started reservations and are making progress toward achieving the goal.

Benchmarking between teams fosters healthy competition to achieve FinOps' goals. The teams who achieve the highest goals are rewarded and the bottom ones are asked to learn from the success of others.

Another tool that FinOps has at its disposal is to establish the budget for each team and allow them to monitor their spending on a daily basis. Let's look at how we can accomplish this using Azure Cost Management.

Creating and managing budgets in Azure cost analysis

Traditionally, budget means setting financial restrictions on how much money a project can spend. But in FinOps language, budget has a positive meaning. In FinOps, you set a cloud spending budget to focus the investment on important innovations. Since the engineers are responsible for creating resources in clouds that cost money, setting a budget at the application level and the department level makes the most sense. Budgets are targets or goals given to product teams as a guard rail and having visibility of how much spend is justifiable to meet business goals generally results in optimal cloud spend. If you are starting the FinOps journey, then you will see no budget set for teams. Your goal as a next step is to work with Finance, Business, and Engineering to establish a reasonable budget. There are various ways you can establish a baseline budget. Organizations in the crawl stage of the FinOps lifecycle might choose to carry forward last year's spending and set it as a budget. Alternatively, you can set a specific budget with a cost takeout goal, for example, looking at last year's spend and reducing it by some percent or amount.

In this section, we will create a budget for our marketing website application, a budget for all dev and production environments, and an overall budget for the marketing department, which includes all applications and environments. The finance team has provided the following information and has asked the FinOps team to set up the budgets accordingly:

Name	Budget	Frequency
Production marketing website budget (dept: marketing and env: prod)	$10,000	Monthly
Marketing development budget (dept: marketing and env: dev)	$500,000	Quarterly
Marketing production budget (dept: marketing and env: prod)	$1,000,000	Quarterly
Overall marketing department budget (dept: marketing)	$18,000,000	Annual

Table 2.2 – Setting up a budget

Let's create our first budget using the Azure cost analysis tool in Azure Portal.

Production marketing website budget

Perform the following steps:

1. Open **Cost Management + Billing** in Azure Portal. Click on **Cost Analysis** and then click on **Budgets**.

> **Note**
> **Budgets** is only available for Enterprise agreements, Microsoft customer agreements, and individual agreements. It is not available for free tier Azure subscriptions.

2. Click on **+ Add** to create our marketing website monthly budget:

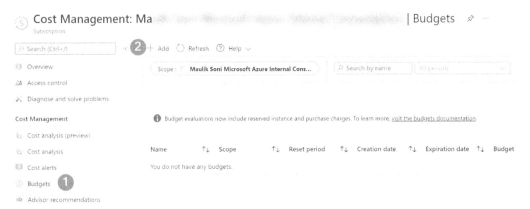

Figure 2.4 – Adding a budget

3. Next, in the **Budget Scoping** section, click on **Add Filters** and add the **Tag** filters of **env= prod** and **dept= marketing**. Then, in the **Budget Details** section, give the budget a name and select **Reset period** as **Monthly**. It has other options such as **Quarterly** and **Annually**, which we will use later. I have set the budget to expire in 2025. Lastly, set the budget amount threshold to **$10,000** as provided by the Finance team:

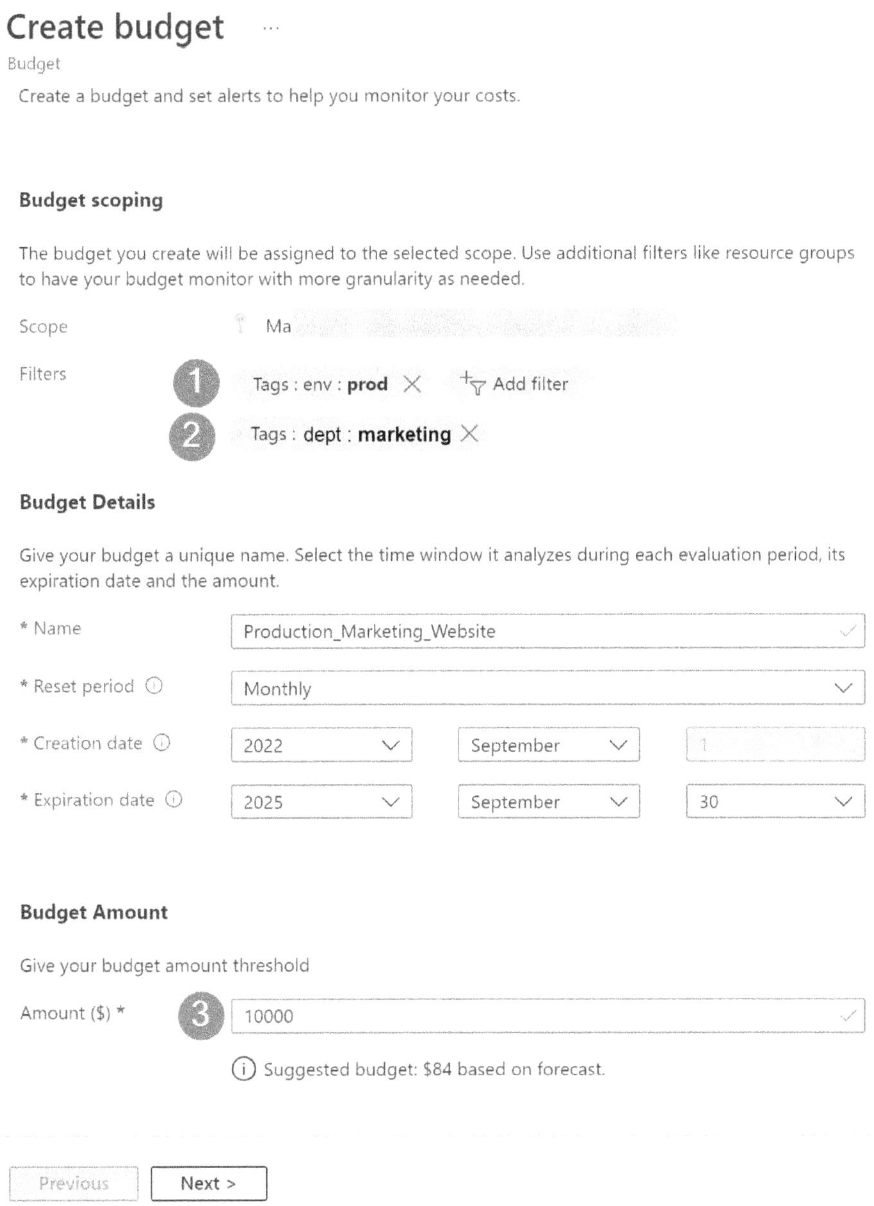

Create budget   ...
Budget
Create a budget and set alerts to help you monitor your costs.

Budget scoping

The budget you create will be assigned to the selected scope. Use additional filters like resource groups to have your budget monitor with more granularity as needed.

Scope	🔑 Ma
Filters	① Tags : env : **prod** ✕ ＋▽ Add filter
	② Tags : dept : **marketing** ✕

Budget Details

Give your budget a unique name. Select the time window it analyzes during each evaluation period, its expiration date and the amount.

* Name	Production_Marketing_Website		✓
* Reset period ⓘ	Monthly		⌄
* Creation date ⓘ	2022 ⌄	September ⌄	1
* Expiration date ⓘ	2025 ⌄	September ⌄	30 ⌄

Budget Amount

Give your budget amount threshold

Amount ($) *	③	10000 ✓

ⓘ Suggested budget: $84 based on forecast.

Previous   Next >

Figure 2.5 – Creating a budget

4.   Click on **Next**. It will now ask you to set up alerts for the budget. Since we have separate sections to discuss alerting patterns, for now, fill in the following details:

I.   Set **Alert conditions** to **Actual** and **90%**.

II.   Provide an alert recipient email and click on **Create**:

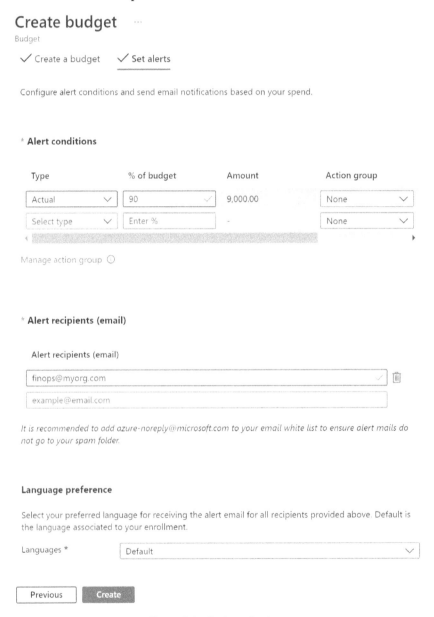

Figure 2.6 – Saving a budget

5. You will be redirected to Budget's main screen and be able to view the budget you have created.

Marketing development budget

Perform the following steps:

1. Click on + **Add** to create our marketing development quarterly budget.

2. Next, in the **Budget Scoping** section, click on **Add Filter** and select the **Tag** options of **env: dev** and **dept: marketing**. Then, in the **Budget Details,** section give the budget a name and select **Reset period** to **Quarterly**. Set the budget to expire in **2025**. Lastly, set the budget amount threshold to **$500,000** as provided by the finance team.

3. Click on **Next** and set up the default alert described in *Step 4* of the *Production marketing website budget* section.

Marketing production budget

Perform the following steps:

1. Click on + **Add** to create our marketing production quarterly budget.

2. Next, in the **Budget Scoping** section, click on **Add Filter** and select **Tag: env: prod** and **dept: marketing**. Then, in the **Budget Details** section, give the budget a name and select **Reset period** as **Quarterly**. Set the budget to expire in **2025**. Lastly, set the budget amount threshold to **$1,000,000**, as provided by the Finance team.

3. Click on **Next** and set up the default alert.

Overall Marketing department budget

Perform the following steps:

1. Click on + **Add** to create our overall Marketing department annual budget.

2. Next, in the **Budget Scoping** section, click on the **Add Tag** filter and apply **dept: marketing**. Then, in the **Budget Details** section, give the budget a name and select **Reset period** to **Annually**. Set the budget to expire in **2025**. Lastly, set the budget amount threshold to **$18,000,000**, as provided by the finance team.

3. Click on **Next** and set up the default alert.

Tracking the budget spend

Once all the budgets have been created, you can go back to the main budget page to view the summary. Here, you can also see the visual progress of the budget, evaluated spend, and progress. Once 100% of the budget is reached, you will see that the progress indicator turns red. At this point, you can use automation if you do not want to go over budget:

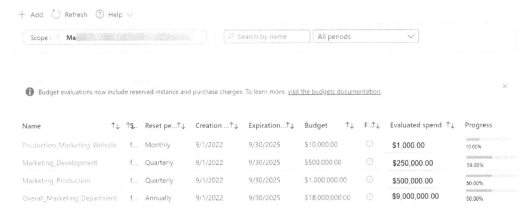

Figure 2.7 – Tracking your budget

Now that we have a budget in place, let's look at alerting strategy and possible automation actions.

Creating and managing alerts in Azure cost analysis

The Azure cost analysis *Budget* feature provides the ability to create and manage cost alerts and anomaly alerts.

> **Note**
>
> Budgets are now evaluated against a more complete set of your costs and include reserved instance and purchase data. If you want to filter the new costs so that budgets are evaluated against first-party Azure consumption charges only, then add **Publisher Type: Azure** and **Charge Type: Usage filter**. When the budget thresholds are exceeded, only notifications are triggered. This does not impact your resources and does not stop consumption.

Let's explore budget alerts in a bit more detail.

Budget alerts

Budget alerts are a capability that you can use to monitor cloud spend and take appropriate actions if there is a deviation in spend. It is important to remember that cost and usage data is generally available between 8–24 hours and budgets are evaluated against these costs every 24 hours. When a budget threshold is met, email notifications are normally sent within an hour of the evaluation.

In the previous section, we created a budget for the application, environment, and department. The finance team has asked us to create the following alerts for each budget:

Name	Type	% of Budget
Half Budget Used	Actual	50%
All Budget Used	Actual	100%

Table 2.3 – Setting up budget alerts

Let's set up the budget alerts using Azure Portal.

Setting up budget alerts

1. Open **Cost Management + Billing** in Azure Portal and then click on **Cost Analysis** and then on **Cost Alerts**.

2. Click on **Manage** and select **Manage Budget** to view all the budgets we created in the previous section.

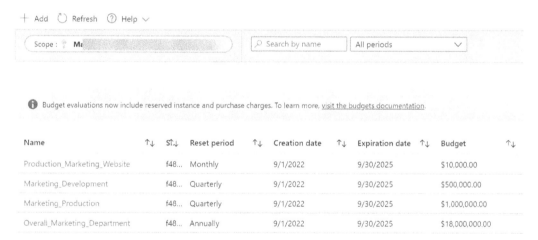

Figure 2.8 – Managing your budget

3. Click on the **Production_Marketing_Website** budget from the list and then click on **Edit Budget**.

4. On the **Edit budget** screen, click on **Set alerts** and set up **50%** threshold and **100%** threshold alerts. Then, click on **Save**:

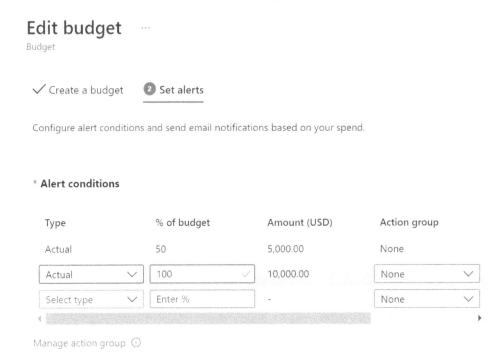

Figure 2.9 – Edit your budget

5. Repeat these steps to create the **Marketing_Development** budget, the **Marketing_Production** budget, and the **Overall_Marketing_Department** budget.

Now that we have the budget and alerts set up, let's configure spending anomaly alerts next.

Spending anomaly alerts

Developers use Terraform, AZ CLI, and other tools to automate resource provisioning in the cloud. Therefore, cloud spend is accumulated from different teams and different projects, and tracking human or tool errors becomes very difficult. This is when you want to create spend anomaly alerts to get notified when unusual changes in resource group costs are detected. An alert will send email notifications to all recipients when an anomaly is detected. Additionally, the notification includes the top resource group changes for the day compared to the previous 60 days.

To add an anomaly alert, perform the following steps:

1. Open **Cost Management + Billing** in Azure Portal. Click on **Cost Analysis** and then click on **Cost Alerts**.

2. Click on +**Add** and select **Add Anomaly Alert** to create a new alert.

3. Provide a name, email subject, and recipients and an optional message.

4. Click on **Save** to create the alert.

The following is an example of a notification email sent by an anomaly alert that we set up.

 Microsoft

Anomaly alert: An unusual cost decrease was detected

An unusual cost decrease was detected on January 18, 2023 for the Ma
subscription. Cost Management detected a possible cost anomaly based on daily cost trends between November 20, 2022 and January 18, 2023. Please review changes to determine whether this was expected.

Subscription summary

Anomaly detected	Yes
Delta compared to expected range	-87.1 %

Resource group summary

- Cost down 640.24% from 9 removed[1] resource group(s).
- Cost changed -26.08% from 16 existing resource group(s).

[1]"Removed" means there was no cost or usage generated. This may be due to stopped, moved, or deleted resources.

Most significant changes in resource group(s) during this period

Name	Cost change %	Percent of total
costmgmt-handbook	-20.93	21.68

Figure 2.10 – Email alert showing details of cost anomaly

The cost anomaly email alert shows the cost increase or decrease by resource group. Cost is estimated based on normalized usage, which standardizes the unit of measure across all usage types (such as hours and GB) and doesn't factor in credits or discounts.

> **Tip**
>
> Check out this Microsoft documentation to manually find an anomaly and what action to take once you receive a notification: `https://docs.microsoft.com/en-us/azure/cost-management-billing/understand/analyze-unexpected-charges`.

This concludes the chapter benchmarking current spend and establishing a budget.

Summary

Here is a recap of what we learned in this chapter. We looked at the on-demand nature of Azure. Remember Azure is a pay-as-you-consume model. If you don't consume, you don't pay. Then, we looked at how to develop KPIs for your business needs. We looked at five attributes of effective KPIs and went over three example KPIs that can be leveraged at your organization. Once you have KPIs developed, you want to create a baseline and start benchmarking teams against those KPIs and business metrics. The scorecard report will be used to encourage competition between teams to achieve the desired results. And finally, we created a budget using the Azure cost analysis tool and set up a cost anomaly alert for the FinOps team to keep an eye on changes in the cloud environment.

In the next chapter, we will explore various ways to forecast future spend and demonstrate how Finance and Procurement can incorporate this forecasting in their business reporting.

3
Forecasting the Future Spend

Financial forecasting estimates future financial outcomes using historical data. FinOps teams are tasked with providing near-accurate data on past cloud service usages and charges so that the finance team can incorporate it in their annual spend forecasting exercise. Forecasting allows the finance team to properly allocate the budget for IT. Typically, forecasting is done annually, but for more mature organizations, it is regularly updated monthly or quarterly when there is a change in operations, business plans, or economic conditions. There are various ways in Azure to obtain past and current usage and charges for such services.

The usage dataset can be extremely large and requires proper solutions to obtain, store, model, and query the data. Forecasting based on past usage can be accomplished by Power BI's native forecasting capabilities, while advanced forecasting by application or business unit requires extra steps to prepare and model the data. Tagging resources is extremely important, and if your historical data does not have tags, then you have to rely on resource group names or subscription names to group costs.

In this chapter, we're going to cover the following main topics:

- Introduction to forecasting
- Getting your Azure usage data
- Setting up the Cost Management connector for Power BI
- Forecasting based on manual estimates
- Forecasting based on past usage
- Advanced forecasting by application
- Fully loaded cost forecasting

Let's get started!

Technical requirements

We will be using the following tools to accomplish the tasks in this chapter:

- The Microsoft Cost Management + Billing tool is available at `https://portal.azure.com/#view/Microsoft_Azure_CostManagement/Menu/~/overview`. Alternatively, you can also find **Cost Management + Billing** by signing into the Azure Portal and, from the top center search bar, typing `Cost Management + Billing`.

- The Microsoft Power BI Cost Management connector is available at `https://learn.microsoft.com/en-us/power-bi/connect-data/desktop-connect-azure-cost-management`.

- DAX Studio is available at `https://daxstudio.org`.

- The SQL Server database with read and write permissions.

- The Azure pricing calculator is available at `https://azure.microsoft.com/en-us/pricing/calculator`.

When using these tools, sign in to the Azure Portal using your organization ID. Please refer to *Chapter 1* for the minimum RBAC permissions required to use **Azure Cost Management + Billing**.

Introduction to forecasting

In accounting, forecasting is the process of using current and historic cost data to predict future costs. It is important to estimate and plan for costs prior to incurring them.

Here is an income statement example of Peopledrift, Inc. to demonstrate forecasting line items:

Forecasted		
Income Statement	**2021**	**2022**
Sales	$ 800.00	$ $960.00
Cost of Sales (COGS)	$ 520.00	$ 624.00
Gross Profit	$ 280.00	$ 336.00
SGA Expense	$ 100.00	$ 120.00
Depreciation	$ 20.00	$ 24.00
EBIT	$ 160.00	$ 192.00
Interest Expense	$ 30.00	$ 37.00
EBT	$ 130.00	$155.00
Taxes	$ 35.00	$ 42.00

Forecasted		
Income Statement	**2021**	**2022**
Net Income	$ 95.00	$ 113.00
Dividends	$ 70.00	$ 83.00
Annual Sales Growth		20.00 %

Table 3.1 – Income statement showing forecasting for 2022

The cloud bill is categorized under **Cost of Goods Sold** (**COGS**), and this is where the finance team will be asking you to provide the forecasted value for the upcoming year.

In the next section, we will show you how to forecast your cloud usage.

Getting your Azure usage data

For the purpose of forecasting, we will need past and current usage data from Microsoft Azure. Apart from that, we will also need the budget, price sheet, and reservation usage details. There are many ways in which we can get this data from Azure:

- **EA portal** – Go to **Reports**, then select **Monthly Report Download**. You can then download Balance and Charges, Usage Detail, Marketplace Charges, and Price Sheet.

- **Azure portal** – Go to **Cost Management** and select **Usage + Charges**. You can download all charges (usage and purchases) and an amortized charges report.

- **Continuous export** – Go to **Cost Management** and select **Exports**. You can specify a storage account and a container in which Azure will export the **Usage and Charges** CSV file. The data can be exported daily, weekly, monthly, or one time.

- **Azure Consumption APIs** – Using programmatic access, you can download usage, marketplace, and reservation details alongside price sheet, budget, and balance data. You can review the documentation for Consumption APIs at `https://learn.microsoft.com/en-us/azure/cost-management-billing/manage/consumption-api-overview`.

- **Azure Cost Management connector for Power BI Desktop** – Using this built-in connector not only gets you the usage data but it also uses Power BI to create custom visualizations and reports based on the usage and charges.

As you can see, there are many ways to get the data we need. Based on your unique scenario, you can obtain the usage data from any of these sources. For the purpose of this chapter, we will be using the Cost Management connector for Power BI for two main reasons. First, it's a low-code solution, and second, we don't need separate analytics/visualization tools since Power BI accomplishes both goals.

Setting up the Cost Management connector for Power BI

Let's review the architecture diagram for our solution. We will be using Power BI to import usage data and transform data, and to create forecasting reports:

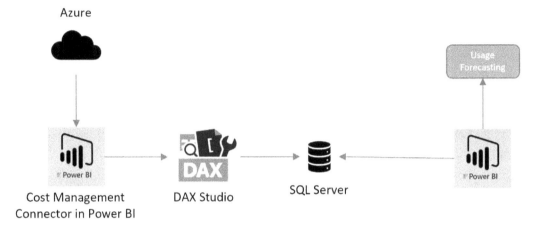

Figure 3.1 – Forecasting solution architecture

Using the Cost Management connector in Power BI, we will query the Azure usage data for the past 12 months. Due to Power BI's limitation of 1 million rows, we will split the import operation into 12 units. Each time we query one month of usage data, we will use DAX Studio to export it from Power BI to the SQL Server database. In the real world, usage files will have millions of records and SQL Server is most suitable to handle that amount of data. Once we have 12 months of data stored in SQL Server, we can start data modeling and forecasting.

Before we begin, please go through the *Technical requirements* section, and make sure you have Power BI Desktop, DAX Studio, and a SQL Server instance ready for use:

1. Open the **Microsoft Power BI Desktop** tool.

2. Under the **Home** tab, click on **Get Data**.

3. Then, select **Azure | Azure Cost Management**:

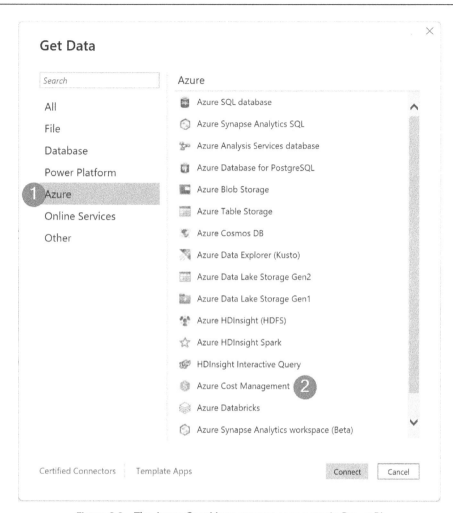

Figure 3.2 – The Azure Cost Management connector in Power BI

4. Next, select **Enrollment Number** as the scope and provide your EA enrollment ID in **Scope Identifier**. Also, it is important to note that when you are selecting the Advanced Options start and end dates, you want to put **Number of months** as **0**. Then, expand **Advanced Options** and select the **Start Date** and **End Date** values. Here, we are selecting September 1, 2021, as the start date and September 30, 2021, as the end date. Remember to select only one month at a time:

Figure 3.3 – Specifying the scope, enrollment number, and start and end dates

5.   Next, select the **Usage details** table and click on **Load**:

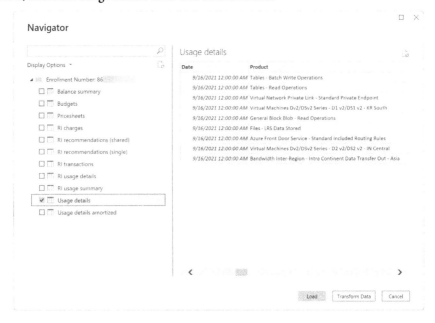

Figure 3.4 – The Usage details table

6. At this point, we have one month of usage data in Power BI. Click on the data view in Power BI and glance at the usage data. At this point, go ahead and save the Power BI report.

7. Next, we need to export this data into SQL Server using DAX Studio. Open DAX Studio from the **Start** menu while keeping the Power BI report open. Then, select **Power BI / SSDT Model**, select your report name, and click on **Connect**:

Figure 3.5 – Connecting DAX Studio to Power BI report

8. Then, in DAX Studio, select the **Advanced** tab, select **Export Data**, and then select **SQL Tables**. Next, specify your SQL Server connection and click on **Next**:

Figure 3.6 – Exporting Power BI data to SQL Server

9. On the next screen, click on **Export** to start exporting one month's worth of data (September 2021) into the SQL Server table.

10. Once the export has been successfully completed, go to your SQL server database instance and use the `select` query for the `[dbo].[Usage details]` table and verify that your cloud usage data has been imported.

Now we have completed a one-month-of-usage data export to SQL Server. Repeat these steps for the rest of the month until you have 12 months of usage data.

> **Note**
>
> As you repeat the steps, make sure to uncheck the **Re-create tables before inserting data** option for next month's data to prevent overwriting. See *Figure 3.6* for reference.

Once you have 12 months of usage data loaded in SQL Server, you are ready to move on to the next steps to understand various forecasting methods.

Forecasting based on manual estimates

Forecasting the cloud spend based on manual estimates is the easiest thing to do and many organizations with a data center mindset use this method of forecasting. There is nothing wrong with using this method as a starting point. The key thing is to evolve to more advanced and accurate forecasting methods as the organization matures.

Let's take an example. The finance team has asked IT to provide manual estimates for IT spending. Historically, IT has tracked spending in two buckets: data center cost and vendor cost. The data center cost represents the cost of provisioned capacity and is revised every three years when the company procures new hardware. Vendor cost is estimated every year based on annual business care reviews and how much outside help is needed to achieve business requirements.

The IT team has provided the following manual estimates for 2023 forecasting:

Manual forecasting	2022	2023
Data center cost	$ 180,000	$ 360,000
Vendor cost	$ 250,000	$ 500,000
Total	$ 430,000	$ 860,000

Table 3.2 – The manual forecasting of IT spend

The annual forecasting is then spread evenly to get the monthly spend. In our example, for 2023, the forecasted data center cost per month is 360,000/12 = $30,000. Finance and IT tracks the monthly spend, and if it is below $30,000, then its *business as usual*. When the spend is higher than the allocated cost, IT needs to provide justification for the deviation or the spike.

Now, let's take a scenario where an organization is already using the Azure cloud to host its infrastructure, but it is beginning its journey in the cloud. To keep things simple, it has decided to do manual forecasting for cloud spend. In this case, Finance took the following approach. They asked IT to provide the total number of VMs and the total number of PaaS databases and then use the Azure pricing calculator to price them. This method is also called cost driver-based forecasting since the two main drivers of cloud costs are VMs and databases:

Item	Quantity	Price calculated using Azure pricing calculator
Total VMs	650	$ 180,000/month
Total PaaS databases	200	$ 75,000/month
Total	850	$ 255,000/month

Table 3.3 – The manual forecasting of cloud spend

As you can see, this method of forecasting cloud spend is not very accurate but simple enough to start the journey of forecasting. In the following sections, we will look at slightly more advanced methods of forecasting the cloud bill.

Forecasting based on past usage

Cloud spend forecasting based on past usage is achieved by querying the usage data and using some form of forecasting algorithm to predict future spend. In our case, the data source will be SQL Server in which we ingested the past 12 months of usage data. For more details, please see the *Getting your Azure usage data* section. We will continue to use Power BI to query the data and create forecasts:

1. Open the **Microsoft Power BI Desktop** tool.

2. Under the **Home** tab in the top toolbar, select **SQL**. Then, select **SQL Server** and connect to your SQL server database where we have imported past usage data. Be sure to select **Data Connectivity mode** as **DirectQuery**:

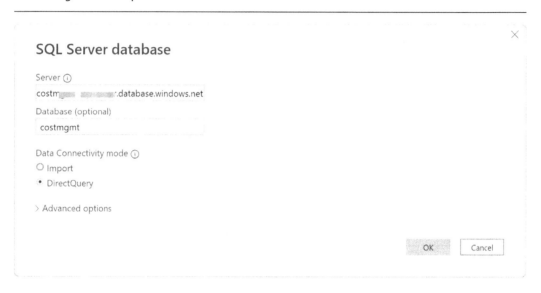

Figure 3.7 – Connecting to SQL Server

3. Next, select the **Usage details** table and click on **Load**:

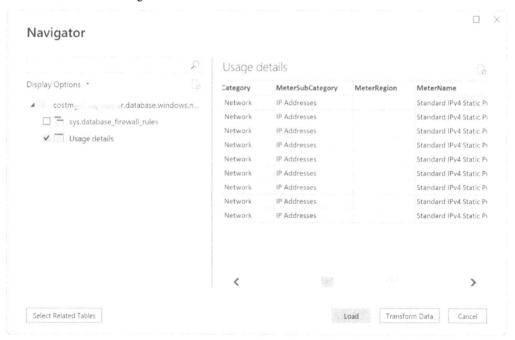

Figure 3.8 – The Usage details table

4. Go to **Report view** in Power BI and insert **Line Chart** from the **Visualizations** section.

5. Drop **Cost** into the *y*-axis and **Billing Period End Date** into the *x*-axis:

Figure 3.9 – Selecting the line chart's X- and Y- axis values

6. Next, we will turn on Power BI's native forecasting. Click on the magnifying glass icon under the **Visualizations** section and turn the **Forecast** toggle to **On**:

Figure 3.10 – Turning on Forecast under the Visualizations section

Once you have turned on the forecasting, you will see the line chart be updated with upper bound and lower bound values:

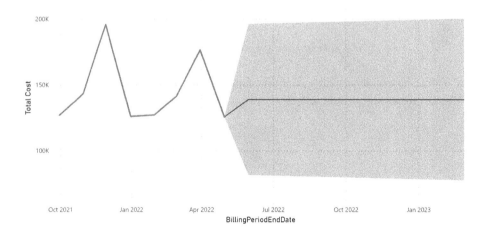

Figure 3.11 – Line chart showing the next 12 months of forecast

7. At this point, as you hover the mouse over the gray area, you will see the Upper bound and lower bound values. We can provide these values back to finance, so they can incorporate them in the planning.

8. Go ahead and save the Power BI report as *Forecasting Based on Past Usage*. We will reuse the data source configured in this report for the next exercise.

Power View performs advanced statistical analysis on your line chart to generate a forecast that incorporates trend and seasonal factors. Power View uses a well-established time-series prediction model called **exponential smoothing**. It uses two versions of exponential smoothing, one for seasonal data (ETS AAA) and one for non-seasonal data (ETS AAN). Power View uses the appropriate model automatically when you start a forecast for your line chart, based on an analysis of the historical data. For more information, please take a look at `https://powerbi.microsoft.com/pt-br/blog/describing-the-forecasting-models-in-power-view/`.

> **Note**
>
> Usage data from the Power BI Cost Management connector includes reservations and Marketplace charges. Reforecasting will be required periodically to make sure new and/or expired reservations are factored into the calculations.

While overall usage-based forecasting is better than manual forecasting, it is not without trade-offs. One of the biggest trade-offs is that the cloud spend is variable. This means Engineering can start and scale workloads without going to Procurement. This ability to create and scale resources at will poses forecasting challenges. To overcome this challenge, we must implement advanced forecasting by service with daily tracking.

Advanced forecasting by application

Organizations with mature FinOps practices want to push the forecasting even further to forecast by application. The idea is that each Engineering team or business unit has a dashboard where they can track the charges incurred by their application. This provides granular visibility to the IT teams and provides an opportunity to track their own spending.

Identifying usage charges by application

One way of mapping service charges back to the application is using *tags*. If your organization is following tagging policy and each resource is tagged, then you can use the **Tags** field in the **Usage Data** table. In our case, the organization does not have sufficient tags in place for historic data and wanted an alternative solution. In that case, the solution is to create a table in Power BI or SQL Database that maps *Subscription ID* to *Application Name*. This can also be easily changed to map resource groups or individual service names to the application if you need further granularity.

Let's implement a solution:

1. Open the **Microsoft Power BI Desktop** tool.

2. Go to the **File** menu and open the **Forecasting Based on Past Usage** report. Since that report already has our data source configured, let's add a new page to the report by clicking on the plus (+) sign at the bottom.

3. Go to **Models view** in Power BI.

4. Under the **Home** tab, select the **Enter Data** button. This allows us to create a table in Power BI. Rename the first column to **SubscriptionID** and add a new column called **ApplicationName**. In the **Name** field, type SubscriptionMapping and click on **Load**:

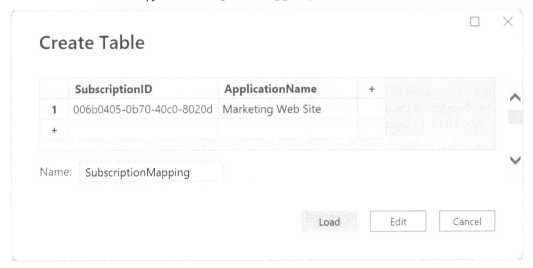

Figure 3.12 – The SubscriptionMapping table in Power BI

5. Next, we need to establish a relationship with the **Usage details** table. Let's do that. In the **Data Model** view, drag the **SubscriptionID** column from the **SubscriptionMapping** table to the **SubscriptionID** column of **Usage details**:

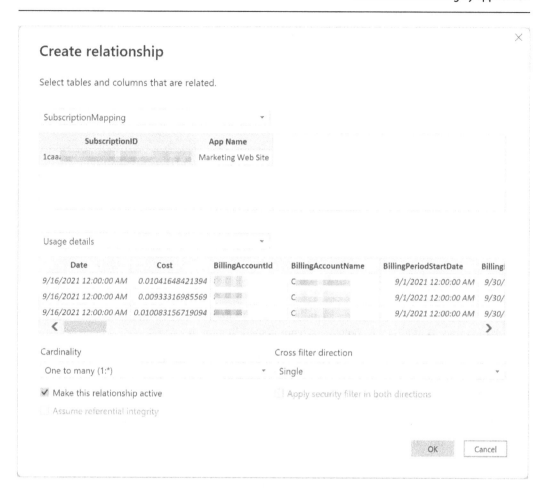

Figure 3.13 – Creating a relationship between tables

6. Once the relationship is in place, go to **Reports View** and add a line chart from the **Visualizations** section.

7. Next, drag **Billing Period End date** on the *x*-axis and **Cost** on the *y*-axis. Then, check the box for the subscription name in the fields.

8. Now, in the **Filter** section, go to the **Subscription Name** field and select the subscription that is mapped to the application. Then, turn on **Forecast** under the **Analytics** section:

Marketing Website Forecasting

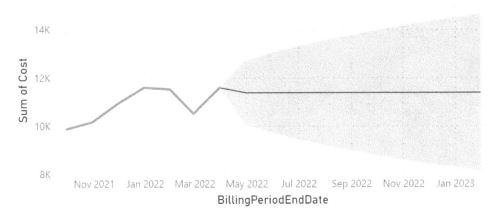

Figure 3.14 – Forecasting by application

Ideally, you might want to create a Power BI dashboard with multiple forecasting charts by applications and track the progress of how the application team is spending money.

Next, let's take a look at how to calculate the fully loaded costs for cloud usage.

Fully loaded costs in forecasting

Fully loaded costs are amortized and include discounts an organization receives from cloud providers using various rate optimization strategies such as the Enterprise Agreement discount, the MACC discount, or even reservations.

The good news is that Azure Cost Management usage data already reflects the fully loaded costs, although it does not include credits, taxes, and other charges. The finance team has now been asked to provide a report displaying reservation charges and Marketplace charges from the usage data. Let's see how we can accomplish this:

1. Open the **Microsoft Power BI Desktop** tool.

2. Go to the **File** menu and open the **Forecasting Based on Past Usage** report. Since that report already has our data source configured, let's add a new page to the report by clicking on the plus (+) sign at the bottom.

3. Insert a table from the **Visualizations** section.

4. Select the table and then select **Cost Center**, **Reservation Name**, and the **Cost** column from the **Fields** section.

5. At this point, we have usage data that includes the reservation and non-reservation costs. We need to filter it down to show only charges where the reservation ID or reservation name is not blank.

6. Select the table, go to the *Filters* section, and check **Select all**, but remove the first item where it's blank, as shown in the following screenshot. This filters out the non-reservation charges:

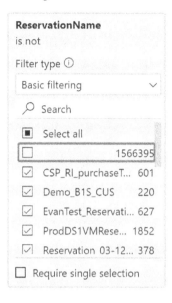

Figure 3.15 – Filter to remove the non-reservation charges

7. Now the table will show your total reservation charges. At this point, you can include the Billing period end date and further filter the report for a specific month, quarter, or year:

CostCenter	ReservationName	Reservation Charges	BillingPeriodEndDate
	Demo_B1S_CUS	20.00	4/30/2022 12:00:00 AM
	Reservation_03-12-2018_14-20	20.00	4/30/2022 12:00:00 AM
	Test-20170919152823	20.00	4/30/2022 12:00:00 AM
	transfertest4	20.00	4/30/2022 12:00:00 AM
	CSP_RI_purchaseTest_03-09-2018_10-59	40.00	4/30/2022 12:00:00 AM
	TransferPatchTest	40.00	4/30/2022 12:00:00 AM
10047774	VM_RI_11-04-2019_14-11	40.00	4/30/2022 12:00:00 AM
	EvanTest_Reservation_03-09-2018_14-54	60.00	4/30/2022 12:00:00 AM
	ProdDS1VMReservation	60.00	4/30/2022 12:00:00 AM
	VM_WUS_DS3_Upfront	80.00	4/30/2022 12:00:00 AM
Total		**400.00**	

Figure 3.16 – Report showing the reservation charges for the month of April 2022

8. Let's save the report. Go to the **File** menu and click on **Save**.

Now you have a report that shows the reservation charges. Let's create a similar report for Marketplace charges:

1. Open the **Microsoft Power BI Desktop** tool.

2. Go to the **File** menu and open the **Forecasting Based on Past Usage** report. Since that report already has our data source configured, let's add a new page to the report by clicking on the plus (+) sign at the bottom.

3. Insert a table from the **Visualizations** section.

4. Select the table and then select the **Cost Center**, **Publisher Name**, **Cost**, and **Publisher Type** columns from the **Fields** section.

5. At this point, we have usage data that includes the Marketplace and Azure services costs. We need to filter it to show only charges where **PublisherType** is **Marketplace**.

6. Select the table and go to the **Filters** section. Under the **PublisherType** field, select **Marketplace** from the list:

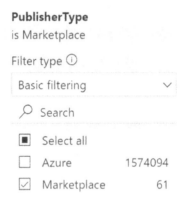

Figure 3.13 – Filter the PublisherType field to Marketplace

7. Now the table in the report will show your total Marketplace charges. At this point, you could include the billing period end date and further filter the report for a specific month, quarter, or year:

CostCenter	PublisherName	Marketplace Purchases	PublisherType
	Citrix	0.00	Marketplace
	Citrix	0.00	Marketplace
	Citrix	0.00	Marketplace
	Citrix	0.00	Marketplace
	Citrix	200.00	Marketplace
P10082818	Cloud Infrastructure Services	4.74	Marketplace
Total		**204.74**	

Figure 3.14 – Report showing Marketplace charges

8. Let's save the report. Go to the **File** menu and click on **Save**.

In this section, we prepared separate reservation and Marketplace reports from the *Usage details* dataset. Let's summarize what we have learned in this chapter.

Summary

We started this chapter with an introduction to forecasting from a cost accounting perspective. We also examined the company's income statement and pinpointed the IT cost categories. Then, we set up our solution and imported usage details data using the Microsoft Power BI Cost Management connector. This connector provides a wealth of usage and charges information. We looked at three methods of forecasting: manual, past usage, and advanced forecasting by application. We ended this chapter by showing how to create separate reservation and Marketplace purchase reports.

Congratulations! You have completed the *Inform* phase of the FinOps lifecycle. What could be more relevant than ending this phase with a case study: let's look at how Peopledrift Inc. started its FinOps journey, the challenges it faced, and the **Objectives and Key Results** (**OKRs**) it achieved.

4

Case Study – Beginning the Azure FinOps Journey

In this chapter, we will review the healthcare company's case study. There is a total of three case studies in this book. Each case study reflects on reinforcing the learning from the previous section. This first case study will focus on Peopledrift Healthcare's journey to build FinOps teams and achieve the goals of the *Inform* phase.

Peopledrift Healthcare came from a very traditional data center mindset until the pandemic hit and the business needed to transform entirely. They started using cloud services to meet customer demands for scheduling appointments and procuring and transporting vaccines. Obviously, this growth was not without pain. It became harder and harder to estimate the current and future spending and Finance was not able to accurately forecast expenditure. To overcome all these challenges, the business implemented FinOps practice and started to improve bit by bit.

In this case study, we will learn about the following:

- Challenges the company faced
- Objectives the company wanted to achieve
- FinOps practice as a solution
- Benefits of FinOps practice

Case study – Peopledrift Healthcare

We are a healthcare company specializing in providing supporting IT services to hospitals and doctors' offices all around the world. Our **Online Appointment Scheduling (OAS)** platform is highly rated by the healthcare industry, and we are the leader in this space. Three years ago, we acquired a small transport and logistics company that powers our *DeliverNow* online platform, which delivers vaccines quickly, with accuracy, and at scale to hospitals, pharmacy stores, and doctors' offices.

Historically, we have owned data centers fully operated by ourselves. We were doing capacity planning 2 years ahead of time and our procurement team would be in intense contract negotiations with lots of different vendors to procure the new hardware to increase the computer, storage, and networking capacity. The IT team provided estimated capacity needs with peak performance and estimated costs to provision the new capacity.

Challenges

Two years ago, a pandemic hit, and we all woke up to a new reality. In just 3 months, demand for our OAS and DeliverNow platform grew 5,000% in terms of **Daily Active Users** (**DAU**) and it kept growing; our data centers were unable to keep up with the need for more hardware. A truly elastic, on-demand, and pay-as-you-use global service was needed to host our infrastructure. This was when our CEO and CTO formulated a strategy to get out of the data center business and move all the workloads to the public cloud within 2 years.

While strategically this was a good decision, our internal processes and teams were not prepared for this digital transformation and needed a framework that could help us transform from the data center to a cloud hosting and operating model.

Objectives

Peopledrift Healthcare has identified the following key objectives for digital transformation:

- No new capital expenditure is to be made on IT infrastructure. Transform people and processes to use the pay-as-you-go model of cloud infrastructure to categorize expenses under Operating Expenditure.

- The company will operate in a hybrid environment until it has been fully migrated to the cloud. Complete migration of both platforms to the cloud by the end of the year. Keep data analytics and business reporting workloads on-premises until next year.

- The cloud cost will be allocated to a single cost center to get the baseline total cost. The cloud bill will be paid using the CTO's special budget.

Solution

To achieve Peopledrift's transformational objectives, we need a framework; a framework that will guide us to start our journey and achieve the short- and long-term strategic goals of moving to the cloud. After careful consideration, the FinOps foundation's FinOps framework was chosen. The FinOps foundation has provided a very carefully crafted roadmap to start and mature the FinOps practice.

Once the framework had been selected, the immediate next step was upskilling the employees who were part of our centralized FinOps team. We prepared a skilling plan as follows and trained two people from the cloud governance team to lead the FinOps practice:

- Take the FinOps Certified Practitioner course from the FinOps foundation: `https://learn.finops.org/path/finops-certified-practitioner-self-paced`

- Learn about Microsoft Azure Fundamentals: `https://learn.microsoft.com/en-us/training/paths/az-900-describe-cloud-concepts`

- Invest time in learning about the Microsoft Azure Well-Architected Framework – Cost Optimization: `https://learn.microsoft.com/en-us/azure/architecture/framework/#cost-optimization`

During the first year, we started with Azure's pay-as-you-go subscription and paid the regular street price for Azure services. By the time the FinOps leads were ready with their knowledge, IT had already started the migration of both our platforms to Azure. The first few subscriptions were created, and IT kept the production and non-production workloads in separate subscriptions. At this point, we were using virtual machines, Application Gateway, Azure App Services, and SQL and CosmosDB databases in Azure.

The first task for the FinOps leads was to gain visibility into the existing environment in Azure. The FinOps leads started arranging the meetings with IT, finance, procurement, and product teams to create the baseline using the Azure Well-Architected Framework's Cost Optimization assessment, available at `https://learn.microsoft.com/en-us/assessments/azure-architecture-review`. This assessment provided us with the starting point of well-defined goals for improvements in terms of CI/CD, disaster recovery, security, and performance, with cost as a primary driver.

The FinOps team then led the effort to organize the Azure account, designed a management group hierarchy, and organized subscriptions under management groups. For management group design, we took the Microsoft recommendations of using a root, platform, landing zone, and sandbox hierarchy. Once the account and management groups were in place, we started seeing the Azure cost organized by department and management groups.

The IT team also updated their processes to include mandatory tags, such as application ID, environment, department, IT owner, business owner, and cost center, in their Infrastructure-as-Code tooling. By the end of the first year, all new resources were created using CI/CD automation with built-in tags. Once the tags were in place, it was easy to create custom Azure cost management dashboards such as Accumulated and Forecasted Cost, Cost by Service, and Cost by Management Groups. This provided much-needed insight into how we spent money in Azure.

Next, the FinOps team wanted to benchmark the cloud expenditure within teams and across other platform teams. Once a use case was established, we wanted to see how two different platform teams utilized IaaS resources and who was better at it. For this purpose, we created autoscale-enabled resources between teams. What we found was that our OAS platform team used Azure virtual machine scale sets with custom autoscaling triggers. This allowed them to dynamically configure the rules so that they could add new virtual machines based on peak demand. Also, lower environments were configured to scale down virtual machines to 0 during the nighttime and weekends. This alone was contributing to 18% savings in the *Compute* category.

The FinOps team, in the last quarter of the first year, worked with finance and defined a strategy to set up a budget for each application in Microsoft Azure. As we were all starting the journey, we used last year's allocated funding and set an annual budget. IT and finance both reviewed the budget utilization on a quarterly basis. This gave finance huge visibility into whether the budget was on track or deviating. In addition to budget alert emails, the Azure cost anomaly alerts were configured.

As the first year passed and the Azure bill was paid by our CTO's special budget, for next year, they wanted the finance team to start forecasting so that the bill would be paid by each business unit. This required the FinOps team to help finance to forecast for the second year. The FinOps team first downloaded and stored historical Azure usage data and used Power BI to create the forecasting. We leveraged Power BI's Azure Cost Management connector to query the enrollment level usage data. One month at a time, we queried the data for the past 12 months and built our SQL Server database usage table. Then, we created a time-series line graph with Power BI's native forecasting feature. This gave us the high and low numbers, which we provided to finance for next year's forecasting.

Benefits

The benefits we received by practicing FinOps were as follows:

- The FinOps team became productive in a very short period of time due to the skilling plan and training efforts.
- The Azure Well-Architected Framework assessment provided awareness of how IT and the business were currently operating in the cloud and helped establish the areas in which we want to improve.
- The finance team was able to see the cost allocated per department, business unit, and application.
- Mandatory tagging was used to aid in cost allocation. The IT team created policies to enforce the tagging, as well as updating their tooling to include the mandatory tags.
- Custom Cost Management reports helped bring additional visibility and provided a mechanism to view costs on a daily basis.
- Cross-team benchmarking presented an opportunity to collaborate on cost-saving activities and provided motivation to each team to be on the leaderboard.
- Setting up a budget in Azure Cost Management per application allowed finance and IT to be on the same page regarding how the budget was being utilized. Automated budget alerts were configured to give early warnings if the budget was expected to overrun.
- Forecasting was accomplished by downloading historical usage data and using Power BI's native forecasting capabilities.

Our FinOps journey does not stop here. Keep reading the next chapter to learn about usage and rate optimization and how we implemented the *Optimize* phase and saved a significant amount of money in monthly bills, which helped us fund new innovative initiatives.

Summary

In this case study, we examined Peopledrift's healthcare service transformation from legacy data center processes and practices to modern FinOps practices to achieve its business, financial, and sustainability goals. The organization was a classic case of businesses having to leverage cloud services during the pandemic to respond to the demand and scale of growing business needs. While that helped business goals, the financial impact was massive. Businesses suddenly realized they were spending a significant amount of money on cloud services, and they needed structured practices to control the cost. Establishing a FinOps practice is just the beginning.

What we have learned so far is foundational. We have visibility of our workload, have tagged the resources and allocated budgets, and have pretty good forecasting reports. The next step is to optimize the current expenditure. Let's look at usage optimization in the next chapter.

Part 2: Optimize

In this part, you will gain an understanding of designing and achieving goals for optimizing cloud service usage and learn techniques for negotiating rates with cloud service providers to reduce costs. The part also covers leveraging strategies to make informed decisions between reservations and savings plans. To solidify your understanding, a case study is presented at the end.

This part contains the following chapters:

- *Chapter 5, Hitting the Goals for Usage Optimization*
- *Chapter 6, Rate Optimization with Discounts and Reservations*
- *Chapter 7, Leveraging Optimization Strategies*
- *Chapter 8, Case Study – Realize Savings and Apply Optimizations*

5
Hitting the Goals for Usage Optimization

In the FinOps Optimize phase, usage optimization is the first opportunity. The reason for that is to pick the low-hanging fruits and generate quick savings with minimum effort. Usage optimization targets fall under two categories: cost avoidance and right sizing. Cost avoidance is achieved by deleting resources that are no longer needed. Using the right sizing strategy, cost savings come by carefully selecting the service SKUs that are just enough for the workload performance needs. We will look at the Iron Triangle method for usage optimization goal settings, and use the **Objectives and Key Results** (**OKRs**) method to identify the cost savings targets. Then, we will use the custom Azure Monitor workbooks to identify the waste and take action to realize the savings by removing or resizing the Azure resources. Azure Advisor also plays a critical role as it automatically monitors the resources and shows right-sizing opportunities. Cost optimization is not without a trade-off, and we will briefly discuss it toward the end of this chapter.

In this chapter, we will learn about the following key topics:

- The project management triangle method for goal setting

- Setting OKRs or KPIs

- Understanding Azure Advisor recommendations for usage optimization

- Top 10 usage optimization targets using custom Azure Workbooks

- Trade-offs of cost versus security, performance, and reliability

Let's get started!

Technical requirements

We will be using the following tools to accomplish the tasks in this chapter:

- Azure Monitor Workbook is available at `https://ms.portal.azure.com/#view/Microsoft_Azure_Monitoring/AzureMonitoringBrowseBlade/~/workbooks/menuId/workbooks`

- Azure Advisor is available at `https://ms.portal.azure.com/#view/Microsoft_Azure_Expert/AdvisorMenuBlade/~/score`

- The Microsoft **Cost Management + Billing** tool is available at `https://portal.azure.com/#view/Microsoft_Azure_CostManagement/Menu/~/overview`. Alternatively, you can also find *Cost Management + Billing* by signing into the Azure portal and, from the top center search bar, typing `Cost Management + Billing`.

When using these tools, sign in to the Azure Portal using your organization ID. Please refer to *Chapter 1* for the minimum RBAC permissions required to use Azure **Cost Management + Billing**.

The project management triangle method for goal setting

The project management triangle, also referred to as the *Iron Triangle* or *Triple constraints*, is a model of constraints that states that changes in one constraint will require changes in other constraints; otherwise, the quality will be impacted. For example, a project can be completed faster by increasing the budget or cutting the scope. Similarly, increasing the scope might require equivalent increases in budget and schedule. Cutting the budget without adjusting the schedule or scope will lead to lower quality:

Figure 5.1 – Project management triangle

So, how is this applicable to Azure cost optimization? This model can be related to cloud decision-making. When a team chooses to reduce costs in their cloud environment, they trade off the quality and speed of the workload.

Quality is measured in the reliability, availability, and consistency of your service. To achieve quality, you must use best-in-class hardware SKUs, top service tiers, and expensive multi-region architecture. If the business requires certain quality attributes for a workload, the engineering teams must project extended costs and time to accomplish the goal.

Time represents the velocity of teams to achieve business goals. The best example is that if the engineering team has less time to develop multi-region read/write database solutions, they have the option to choose CosmosDB over any open source database. This will significantly boost their time to market since IT can leverage readymade solutions to fit their needs. However, this also means they have to compromise on increased costs to achieve speed.

Cost refers to the budgeted amount of IT spend in the cloud to achieve a business goal. If the allocated amount for the cloud is less, then the resulting solution will not have the necessary attributes of quality and speed. Due to limited budgets, IT is restricted to making the best choices for the workload's reliability, security, and availability.

As you can see, changing one axis of a triangle results in adjusting the other two. Next, let's look at how we can set the key objectives or KPIs for usage optimization.

Setting OKRs or KPIs

OKR is a collaborative goal-setting framework used by teams to set transformative goals with measurable results. OKRs are used to track the progress teams are making toward the goal and align and encourage them to achieve the objective. The OKR methodology was created by Andy Grove at Intel. Let's look at some OKR examples next.

OKR examples

When you design the OKRs for your organization, keep in mind that they must be credible, sustainable, and controllable. If the teams do not trust the OKRs, then the OKRs defined by the FinOps teams are not going to achieve the expected results. You create credibility by making things transparent. Each team should have access to the data that makes up the OKRs. OKRs need to be tracked quarterly or semi-annually:

Objective	Key results
Save 25% on batch job processing infrastructure	<ul><li>Right -size the compute SKU using CPU, memory, and network bandwidth metrics</li><li>Implement autoscaling based on demand or the time of the day</li><li>Offload notification processing to consumption-based serverless functions</li></ul>

Objective	Key results
Save 20% on non-production environments	• Terminate expired sandbox subscriptions • Using automation, shut down development and QA infrastructure during the weekend • Use SQL Server database pools to share the resources

Table 5.1 – OKR examples

Having a product owner manage the OKRs for the Engineering team makes more sense than leaving it up to engineering to report on it. Next, let's look at the KPIs.

KPI – tagging by business unit

The tagging by business unit KPI shows the current state of tagging coverage per BU. The KPI also has a goal line that each BU needs to reach by the end of the quarter:

Figure 5.2 – Tagging by business unit KPI

The FinOps team has created this KPI to increase the coverage for mandatory tagging up to the goal line for each business unit.

KPI – cost avoidance for unattached disks by business unit

The cost avoidance KPI tracks the virtual machine disks that are orphaned or, in other words, unattached. Even though the virtual machine is deleted, the disk is still incurring charges:

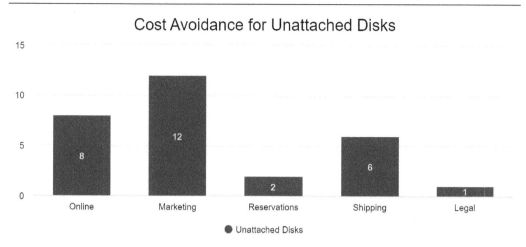

Figure 5.3 – The Cost Avoidance for Unattached Disks KPI

The FinOps team has created this KPI to remove waste by showing the unattached disks by business unit. The goal is that there should be zero unattached disks for all business units.

KPI – Azure Hybrid Benefit utilization by business unit

This KPI tracks the virtual machines without Hybrid Benefit. The organization has hybrid use licenses, but they are not enabled for Azure Virtual Machines.

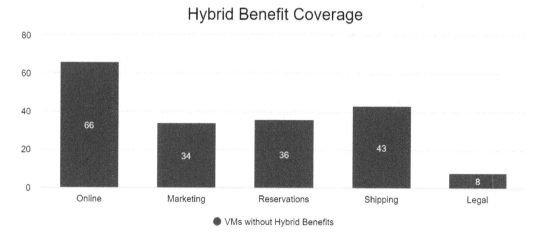

Figure 5.4 – VMs without Hybrid Benefit KPI

The FinOps team keeps track of potential savings opportunity KPIs. By enabling Hybrid Benefit, the business units can gain up to 20–30% in savings. The goal is to have no VMs without Hybrid Benefit. To learn more about Azure Hybrid Benefit, please take a look at the FAQ page at `https://azure.microsoft.com/en-in/pricing/hybrid-benefit/faq`.

KPI – storage accounts with hot, cool, and archive tiers

The FinOps team has created this KPI to increase the use of the archive tier. Across the organization, the engineering team is not utilizing the lowest- cost archive tiers and missing a cost savings opportunity. The goal of this KPI is to decrease the hot tiers and increase cool and archive tier usage:

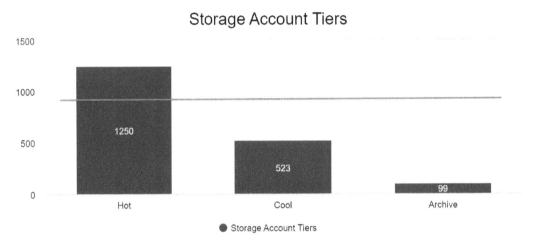

Figure 5.5 – The storage account tiers KPI

Now, let's take a look at Azure Advisor and see what recommendations it offers for our workload.

Understanding Azure Advisor recommendations for usage optimization

Azure Advisor is a free service that examines resources and their usage in your subscriptions and provides personalized recommendations to help you optimize costs. You can configure Azure Advisor to target specific subscriptions or resource groups to selectively focus on critical optimizations. You can access Advisor using the portal, CLI, or REST API.

Here are some example recommendations Azure Advisor can provide to reduce costs:

- Right-size or shut down any underutilized VMs
- Use standard storage to store managed disk snapshots
- Right-size underutilized PostgreSQL servers
- Consider acting on your idle Azure Cosmos DB containers
- Right-size Data Explorer resources for optimal costs
- Repurpose or delete idle virtual network gateways

- Revisit the retention policy for classic log data in storage accounts

- Configure automatic renewal for expiring reservations

- Buy VM reserved instances to save money over pay-as-you-go costs

- Consider an Azure Cosmos DB reserved instance to save over your pay-as-you-go costs

- Consider enabling the autoscaling feature on Spark compute

To see a complete list of Azure Advisor recommendations with detailed explanations, please visit `https://learn.microsoft.com/en-us/azure/advisor/advisor-reference-cost-recommendations`.

Accessing Azure Advisor using the portal

Azure Advisor provides a quick look into savings opportunities you might have with your currently deployed resources. It's an easy first step to look at all the recommendations it presents. Azure Advisor looks back at your resource utilization for up to 60 days and provides personalized recommendations on the right size, an SKU change, or the removal of any resources that are not in use.

To access the Azure Advisor in Azure Portal, perform the following steps:

1. Open the **Microsoft Edge** browser.

2. Navigate to `https://portal.azure.com` and sign in using your organization's account.

3. In the top search bar, search for `Advisor` and select the highlighted service.

4. On the **Overview** page, click on **Cost**:

Figure 5.6 – Azure Advisor displaying recommendations

5. Click on **Recommendation** to see the details:

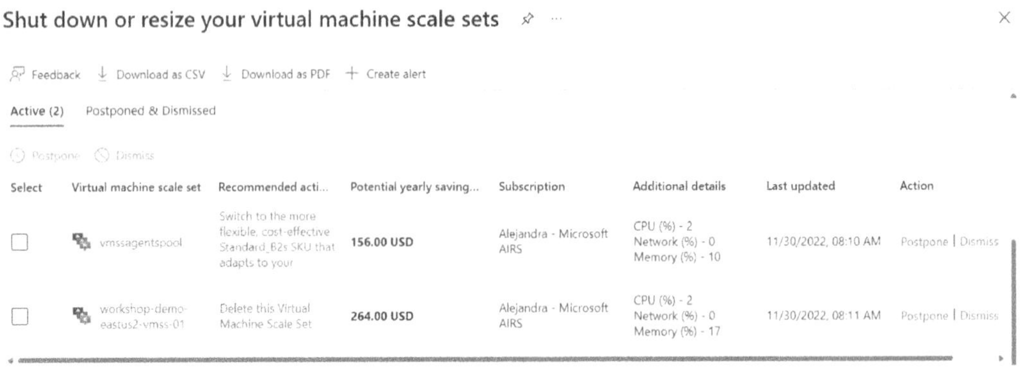

Shut down or resize your virtual machine scale sets ✎ ···   ✕

Figure 5.7 – Azure Advisor recommendation details

6. At this point, you can follow the suggestion by clicking on the **Recommended action** link. If you need more time to act on this suggestion, you can click on **Postpone**. Based on the timeframe you select (between 1 day and 3 months), the suggestion will not show until it's due. To remove this suggestion from the list, you can click on **Dismiss**.

For certain suggestions, you will see the **Quick Fix** tag. This means you can remediate the issue with a single click. Now, let's look at how we can access the advisor using the CLI.

Accessing Azure Advisor using the CLI

If you want programmatic access to Azure Advisor recommendations, use the following CLI command to list the cost-related recommendations:

```
az advisor recommendation list --category Cost -o yaml
```

The command will output the following result:

```
- category: Cost
  extendedProperties:
    MaxCpuP95: '2'
    MaxMemoryP95: '10'
    MaxTotalNetworkP95: '0'
    annualSavingsAmount: '156'
    currentSku: Standard_D2_v3
    deploymentId: 70a30efb-4de0-5394
    recommendationMessage: Switch to the more flexible, cost-
effective Standard_B2s SKU that adapts to your workload
```

```
    recommendationType: SkuChange
    regionId: uswest
    roleName: VMSSAGENTSPOOL
    savingsAmount: '13'
    savingsCurrency: USD
    subscriptionId: a138-47f1-a47a-xxx
    targetSku: Standard_B2s
  id: /subscriptions/...4b65
  impact: High
  impactedField: Microsoft.Compute/virtualMachineScaleSets
  impactedValue: vmssagentspool
  lastUpdated: '2022-11-30T14:10:46.835063+00:00'
  metadata: null
  name: 1678-13fe-9d68a60c4b65
  recommendationTypeId: 94aea435-xxx
  resourceGroup: vmssagents
  resourceMetadata:
    resourceId: /subscriptions/xxx/resourcegroups/vmssagents/
providers/microsoft.compute/virtualmachinescalesets/
vmssagentspool
    source: null
  risk: null
  shortDescription:
    problem: Right-size or shutdown underutilized virtual
machine scale sets
    solution: Right-size or shutdown underutilized virtual
machine scale sets
  suppressionIds: null
  type: Microsoft.Advisor/recommendations
```

Note the key properties the preceding output shows. The recommendationMessage field shows the actual message that Advisor has identified. In the preceding example, Advisor wants us to change the VM SKU from D-Series to a burstable VM based on its utilization telemetry from the past 30 days. The annualSavingsAmount field tells us that by acting on these recommendations, we will save $156 annually. And lastly, the output provides us with the target resource ID to identify the resource that needs to be updated.

Next, let's look at custom workbooks that can help with our usage optimization goals.

Top 10 usage optimization targets using custom Azure workbooks

Usage optimization refers to the activities around removing, right-sizing, and redesigning workloads and services in the cloud with the end goal of saving money. Often, removing means resources in the cloud are created and forgotten. No one knows what they are used for or even whether they are needed for the business to function. Identify those resources and remove them. Right-sizing means scaling down or out by looking at utilization over a period. Often, cloud services are provisioned with their peak in mind, and by looking at CPU, memory, discs, and networking metrics, you can determine that the same performance can be achieved by a different SKU. On the other hand, redesigning the workload means making a fundamental change in how the workload is architected. Often, the redesign brings the best cost savings, but it takes longer and requires development efforts.

Target 1 – 98% of all your resources must be tagged

Tagging plays a pivotal role in allocating and tracking the budget for each application. When following the best practices for tag management, tags can be the basis for applying your business policies with Azure Policy or tracking costs with Cost Management. The goal is that 98% of all Azure resources must have all the required tags. Use **Workbook 1** from Azure Monitor to find tagged versus untagged resource groups and resources. Use the untagged resources list to tag the resources that have missing tags:

1.   Go to Azure Monitor and create a new empty workbook:

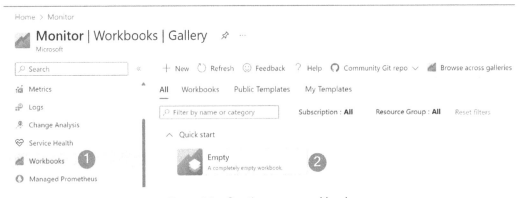

Figure 5.8 – Creating a new workbook

2.   Next, click on **Advanced Editor** and select **Gallery Template**:

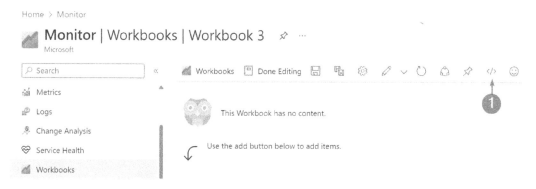

Figure 5.9 – Advanced Editor

3.   Copy and paste the code from `https://github.com/PacktPublishing/FinOps-Handbook-for-Microsoft-Azure/tree/main/Chapter-5` and select `Workbook 1`:

Figure 5.10 – Copying and pasting the workbook code from GitHub

4.   Click on **Apply** and then click on **Done Editing**:

Image 5.11 – Workbook showing tagged versus untagged resources

5.   Set **Workbook Auto Refresh** to **30 minutes**.

6.   The workbook uses the following Azure Resource Graph query. You can certainly apply additional filters and conditions to suit your need:

```
ResourceContainers| where type =~ 'Microsoft.Resources/
subscriptions'
| extend SubscriptionName=name
| join  ( ResourceContainers | where type =~ 'microsoft.
resources/subscriptions/resourcegroups' | where tags =~
'' or tags =~ '{}'
| extend resourceGroupName=id, RGLocation=location) on
subscriptionId
| project resourceGroupName, RGLocation, SubscriptionName
```

Let's move on to the next target.

Target 2 – right-sizing underutilized virtual machines

Right-sizing is a key lever to reduce costs and optimize resources. It's an exercise to find out the smallest VM instance that supports your workload requirements. Azure Advisor automatically evaluates your cloud footprint and analyzes the CPU, memory, and network utilization of the past 7 days. Then, using algorithms, it calculates the appropriate SKU or instance count. Azure Advisor determines the best fit and the cheapest costs with no performance impacts and generates the recommendation for right-sizing.

Use `Workbook 2` as it provides a list of right-size recommendations for your VMs and VM scale sets:

1. Go to **Azure Monitor** and create a new empty workbook.

2. Next, click on **Advanced Editor** and select **Gallery Template**.

3. Copy and paste the code from `https://github.com/PacktPublishing/FinOps-Handbook-for-Microsoft-Azure/tree/main/Chapter-5` and select `Workbook 2`.

4. Click on **Apply** and then click on **Done Editing**:

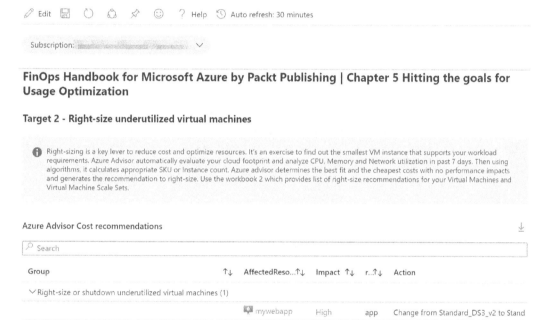

Figure 5.12 – Workbook showing the right-size VM recommendations

5. Set **Workbook Auto Refresh** to **30 minutes**.

6. The workbook uses the following Azure Resource Graph query. You can certainly apply additional filters and conditions to suit your need:

```
advisorresources
| where type == "microsoft.advisor/recommendations"
| where tostring (properties.category) has "Cost"
```

```
| where properties.impactedField has "Compute" or
properties.impactedField has "Container" or properties.
impactedField has "Web"
| project AffectedResource=tostring(properties.
resourceMetadata.resourceId),Impact=properties.
impact,resourceGroup,Action=tostring(properties.extended
Properties.
recommendationMessage),AdditionaInfo=properties.
extendedProperties,subscriptionId,Recommendation=tostring
(properties.shortDescription.problem),name
```

Let's move on to the next target.

Target 3 – enabling Azure Hybrid Benefit for Windows and Linux VMs

VMs deployed from pay-as-you-go images on Azure incur both an infrastructure fee and a software fee. Azure Hybrid Benefit for Windows Server allows you to use your on-premises Windows Server licenses and run Windows VMs on Azure at a reduced cost. Now Azure also offers Bring Your Own License for **Red Hat Enterprise Linux** (**RHEL**) and **SUSE Linux Enterprise Server** (**SLES**) VMs. With this benefit, your RHEL or SLES subscription covers your software fee. You only pay infrastructure costs for your VM. Use **Workbook 3** to list out your Windows and Linux VMs that do not have Hybrid Benefit enabled. Then, use the CLI or Azure Portal to enable Hybrid Benefit to optimize the savings:

1. Go to **Azure Monitor** and create a new empty workbook.

2. Next, click on **Advanced Editor** and select **Gallery Template**.

3. Copy and paste the code from `https://github.com/PacktPublishing/FinOps-Handbook-for-Microsoft-Azure/tree/main/Chapter-5` and select `Workbook 3`.

4. Click on **Apply** and then click on **Done Editing**:

FinOps Handbook for Microsoft Azure by Packt Publishing | Chapter 5 Hitting the goals for Usage Optimization

Target 3 - Enable Azure hybrid benefit for Windows and Linux virtual machines

Windows VMs without Azure Hybrid Benefit enabled

WindowsId	↑↓	VMName	↑↓	VMRG	↑↓	VMSize	↑↓	VMLocation	↑↓	OSType	↑↓	OsVersio
UAE-VM-11		UAE-VM-11		uae-rg-01		Standard_A1_v2		uaenorth		Windows-10		19h2-ent
WEU-Dev-VM-01		WEU-Dev-VM-01		weu-dev-fin		Standard_A1_v2		westeurope		Windows-10		19h2-ent

Linux VMs without Azure Hybrid Benefit enabled

WindowsId	↑↓	VMName	↑↓	VMRG	↑↓	VMLocation ↑↓	OSType ↑↓	OsVersion ↑↓	LicenseType ↑↓	VMSize
northwindriskserver2		northwindriskserver2		northwindrisk		eastus	RHEL	82gen2		Standard

Figure 5.13 – Workbook showing Windows and Linux VMs without Hybrid Benefit

5. Set **Workbook Auto Refresh** to **30 minutes**.

6. To enable Azure Hybrid Benefit, you can use Azure Portal or the PowerShell command. In Azure Portal, go to the VM and then select **Configuration**. Under the **Licensing** option, select **Yes** to use the existing Windows Server license:

Figure 5.14 – Hybrid Benefit using Azure Portal

7. The workbook uses the following Azure Resource Graph query. You can certainly apply additional filters and conditions to suit your needs:

```
ResourceContainers | where type =~ 'Microsoft.Resources/
subscriptions' | extend SubscriptionName=name | join
( resources | where type =~ 'microsoft.compute/
virtualmachines' and  tostring(properties.storageProfile.
osDisk.osType) == 'Windows' and tostring(properties.
['licenseType']) !has 'Windows' | extend WindowsId=id,
VMName=name, VMLocation=location, VMRG=resourceGroup,
OSType=tostring(properties.storageProfile.
imageReference.offer), OsVersion = tostring(properties.
storageProfile.imageReference.sku), VMSize=tostring
(properties.hardwareProfile.vmSize), LicenseType =
tostring(properties.['licenseType'])) on subscriptionId
    | order by VMSize asc
        | project WindowsId,VMName,VMRG,VMSize,
VMLocation,OSType, OsVersion,LicenseType,
SubscriptionName
```

Let's move on to the next target.

Target 4 – right-sizing underutilized SQL databases

Azure Advisor samples the utilization of SQL databases and, based on an assessment, it recommends right-sizing and provides a new **Database Transaction Unit** (**DTU**) and **Stock Keeping Unit** (**SKU**), or if the utilization is 0 DTUs, it will recommend deleting the instance. Low-resource utilization results in unwanted costs, which can be fixed without a significant performance impact. Use **Workbook 4** to see which SQL databases are underutilized and act on Azure Advisor's recommendations:

1. Go to **Azure Monitor** and create a new empty workbook.

2. Next, click on **Advanced Editor** and select **Gallery Template**.

3. Copy and paste the code from `https://github.com/PacktPublishing/FinOps-Handbook-for-Microsoft-Azure/tree/main/Chapter-5` and select `Workbook 4`.

4. Click on **Apply** and then click on **Done Editing**:

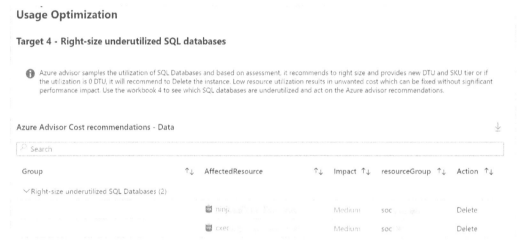

Figure 5.15 – Workbook showing underutilized SQL databases

5. Set **Workbook Auto Refresh** to **30 minutes**.

6. The workbook uses the following Azure Resource Graph query. You can certainly apply additional filters and conditions to suit your need:

```
advisorresources
| where type == "microsoft.advisor/recommendations"
| where tostring (properties.category) has "Cost"
| where properties.impactedField has "Data" or
properties.impactedField has "Sql"
```

```
| project AffectedResource=tostring(properties.
resourceMetadata.resourceId),Impact=properties.
impact,resourceGroup,Action=tostring(properties.
extendedProperties.Recommended_
SKU),AdditionaInfo=properties.
extendedProperties,subscriptionId,Recommendation=tostring
(properties.shortDescription.problem),name
```

Let's move on to the next target.

Target 5 – enabling Azure Hybrid Benefit for SQL databases, managed instances, and SQL VMs

Azure Hybrid Benefit for SQL Server allows us to use SQL Server licenses with Software Assurance to reduce the base rate for SQL Database Managed Instance and SQL VMs. Azure is the only cloud that provides the ability to apply a license to a fully managed PaaS product. Use Azure Hybrid Benefit for SQL Server with Azure Hybrid Benefit for Windows Server to get maximum savings. Use **Workbook 5** to find out the SQL Server instances without Hybrid Benefit. Review the instances and apply Azure Hybrid Benefit to the matching instances:

1. Go to **Azure Monitor** and create a new empty workbook.

2. Next, click on **Advanced Editor** and select **Gallery Template**.

3. Copy and paste the code from `https://github.com/PacktPublishing/FinOps-Handbook-for-Microsoft-Azure/tree/main/Chapter-5` and select `Workbook 5`.

4. Click on **Apply** and then click on **Done Editing**:

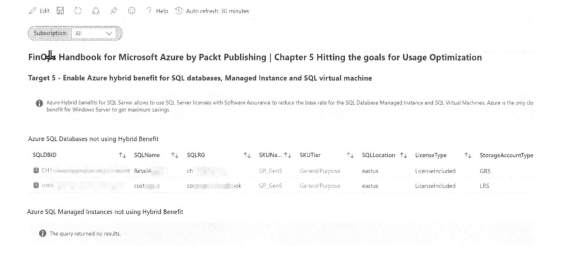

Figure 5.16 – Workbook showing SQL Server instances without Hybrid Benefit

5. Set **Workbook Auto Refresh** to **30 minutes**.

6. The workbook uses the following Azure Resource Graph query. You can certainly apply additional filters and conditions to suit your need:

```
ResourceContainers
| where type =~ 'Microsoft.Resources/subscriptions'
| extend SubscriptionName=name | join (resources | where
type =~ 'Microsoft.Sql/servers/databases' and name !=
'master' and tostring(properties.['licenseType']) ==
'LicenseIncluded'
| extend    SQLDBID=id,SQLName = name, SQLRG =
resourceGroup, SKUName=sku.name, SKUTier=sku.tier,
SQLLocation = location, LicenseType = tostring(properties.
['licenseType']), StorageAccountType=tostring(properties.
['storageAccountType'])) on subscriptionId
| project SQLDBID,SQLName,SQLRG, SKUName, SKUTier,
SQLLocation, LicenseType, StorageAccountType,
SubscriptionName
```

Let's move on to the next target.

Target 6 – upgrading storage accounts to General-purpose v2

General-purpose v2 storage accounts support the latest Azure Storage features and incorporate all the functionality of general-purpose v1 and Blob storage accounts. General-purpose v2 accounts are recommended for most storage scenarios. General-purpose v2 accounts deliver the lowest per-gigabyte capacity prices for Azure Storage, as well as industry-competitive transaction prices. General-purpose v2 accounts support default account access tiers of hot or cool and blob-level tiering of hot, cool, or archive. Use **Workbook 6** to find the v1 storage accounts across the subscriptions. Once you have determined they are safe to upgrade, you can use Azure Portal or the CLI to upgrade your existing v1 storage accounts to v2:

1. Go to **Azure Monitor** and create a new empty workbook.

2. Next, click on **Advanced Editor** and select **Gallery Template**.

3. Copy and paste the code from `https://github.com/PacktPublishing/FinOps-Handbook-for-Microsoft-Azure/tree/main/Chapter-5` and select `Workbook 6`.

4. Click on **Apply** and then click on **Done Editing**:

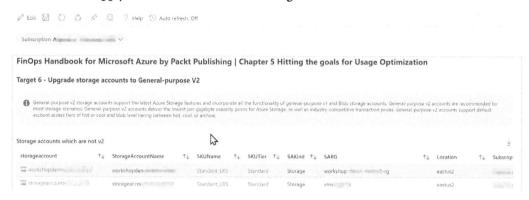

Figure 5.17 – Workbook showing v1 storage accounts

5. Set **Workbook Auto Refresh** to **30 minutes**.

6. The workbook uses the following Azure Resource Graph query. You can certainly apply additional filters and conditions to suit your need:

```
ResourceContainers | where type =~ 'Microsoft.Resources/
subscriptions' | extend SubscriptionName=name |
join (resources | where type =~ 'Microsoft.Storage/
StorageAccounts' and kind !='StorageV2' | extend
storageaccount=id, StorageAccountName=name, SAKind=kind,
SKUName=sku.name, SKUTier=sku.tier, SARG=resourceGroup,
Location=location) on subscriptionId
 | order by id asc
 | project storageaccount,StorageAccountName, SKUName,
SKUTier, SAKind, SARG, Location, SubscriptionName
```

Let's move on to the next target.

Target 7 – deleting unattached discs

When a VM is deleted, the attached discs are preserved. The attached discs' cost is based on the capacity, replication model, and type. The most expensive discs are premium discs with GRS or RA-GRS types. Use **Workbook 7** to find the unattached discs across the subscriptions. Once you have determined they are no longer needed, you can either delete them or move them to archival storage:

1. Go to **Azure Monitor** and create a new empty workbook.

2. Next, click on **Advanced Editor** and select **Gallery Template**.

3. Copy and paste the code from `https://github.com/PacktPublishing/FinOps-Handbook-for-Microsoft-Azure/tree/main/Chapter-5` and select `Workbook 7`.

4. Click on **Apply** and then click on **Done Editing**:

Figure 5.18 – Workbook showing unattached discs

5. Set **Workbook Auto Refresh** to **30 minutes**.

6. The workbook uses the following Azure Resource Graph query. You can certainly apply additional filters and conditions to suit your need:

```
ResourceContainers | where type =~ 'Microsoft.
Resources/subscriptions' | extend SubscriptionName=name
| join (resources | where type =~ 'microsoft.
compute/disks' and tostring (properties.diskState)
== 'Unattached' and name !has '-ASRReplica' |
extend DiskID=id, DiskName=name, SKUName=sku.name,
SKUTier=sku.tier, DiskSizeGB=tostring(properties.
diskSizeGB),DiskRG=resourceGroup, Location=location,
TimeCreated=tostring(properties.timeCreated))on
subscriptionId
| order by id asc
| project DiskID,DiskName, DiskSizeGB, SKUName, SKUTier,
DiskRG, Location, TimeCreated, SubscriptionName
```

Let's move on to the next target.

Target 8 – deleting unattached public IPs

Standard static IP addresses in Azure cost around $4. It is often possible that workloads that are associated with these IP addresses may no longer exist, but the IP address is preserved. Use **Workbook 8** to find the unattached IP addresses across subscriptions. Once you have determined they are no longer needed, delete these IP addresses to save money:

1. Go to **Azure Monitor** and create a new empty workbook.

2. Next, click on **Advanced Editor** and select **Gallery Template**.

3. Copy and paste the code from `https://github.com/PacktPublishing/FinOps-Handbook-for-Microsoft-Azure/tree/main/Chapter-5` and select `Workbook 8`.

4. Click on **Apply** and then click on **Done Editing**:

Figure 5.19 – Workbook showing unattached IPs

5. Set **Workbook Auto Refresh** to **30 minutes**.

6. The workbook uses the following Azure Resource Graph query. You can certainly apply additional filters and conditions to suit your need:

```
ResourceContainers | where type =~ 'Microsoft.
Resources/subscriptions' | extend SubscriptionName=name
| join (resources | where type =~ 'Microsoft.
Network/publicIPAddresses' and isempty(properties.
ipConfiguration) | extend ipdetails=id,
IPName=name, AllocationMethod=tostring(properties.
publicIPAllocationMethod), SKUName=sku.name,
PIPRG=resourceGroup, Location=location)on subscriptionId
| order by ipdetails asc  | project ipdetails,IPName,
SKUName, PIPRG, Location, AllocationMethod,
SubscriptionName
```

Let's move on to the next target.

Target 9 – Azure App Service – using the v3 plan with reservations and autoscaling

Azure App Service has various ways to control and optimize costs. The App Service v3 plan gives 20% discounts over the v2 plan for comparable configurations. The v3 plan also supports reservations. The reservation discount is applied automatically to the number of running instances that match the reservation scope and attributes. Leverage App Service's excellent ability to scale out and scale in. Using autoscaling, you can further optimize the cost by matching your workload requirements for compute capacity. Use **Workbook 9** to identify older v2 App Service plans and upgrade them to v3. Also, verify that autoscaling is enabled for all the web apps:

1. Go to **Azure Monitor** and create a new empty workbook.

2. Next, click on **Advanced Editor** and select **Gallery Template**.

3. Copy and paste the code from `https://github.com/PacktPublishing/FinOps-Handbook-for-Microsoft-Azure/tree/main/Chapter-5` and select `Workbook 9`.

4. Click on **Apply** and then click on **Done Editing**:

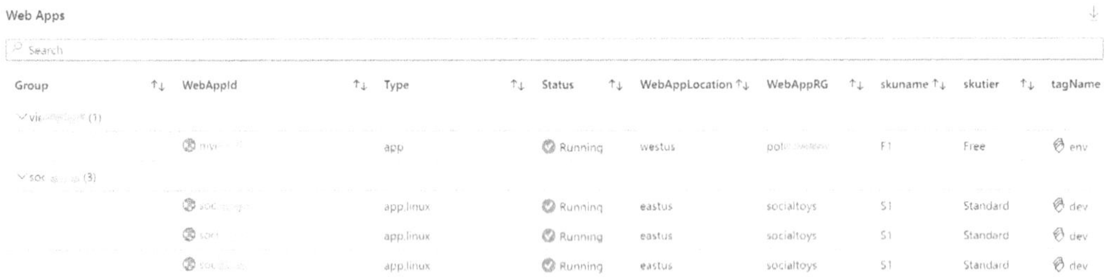

Figure 5.20 – Workbook showing all the App Service plans

5. Set **Workbook Auto Refresh** to **30 minutes**.

6. The workbook uses the following Azure Resource Graph query. You can certainly apply additional filters and conditions to suit your need:

```
ResourceContainers | where type =~ 'Microsoft.Resources/
subscriptions'
| extend SubscriptionName=name
| join( resources | where type =~ 'Microsoft.Web/sites'
| extend WebAppId=id, WebAppRG=resourceGroup,
WebAppName=name, AppServicePlan=tostring(properties.
serverFarmId), SKU=tostring(properties.sku),
Type=kind, Status=tostring(properties.state),
```

```
WebAppLocation=location) on subscriptionId
 | order by WebAppId asc
 | join ( resources | mvexpand tags |
extend  WebAppName=name,Tags=tags, resourceGroup
 | extend tagName = tostring(bag_keys(tags)[0])
 | extend tagValue = tostring(tags[tagName]) ) on
WebAppName
 | project WebAppId,WebAppName, Type,
Status, WebAppLocation, AppServicePlan,
WebAppRG,tagName,tagValue,SubscriptionName
```

Let's move on to the next target.

Target 10 – Azure Kubernetes Service – using the cluster autoscaler, Spot VMs, and start/stop features in AKS

Azure Kubernetes Service (**AKS**) clusters are most common nowadays due to the popularity of containers. With hundreds of clusters in your subscriptions, various cost savings opportunities are available. Use the cluster autoscaler to automatically adjust the number of agent nodes based on workload. Use Spot VMs for workloads that can handle interruptions. Use the horizontal pod autoscaler to adjust the number of pools depending on CPU utilization and other metrics. Finally, use the cluster start/stop feature in AKS for development/test scenarios to reduce costs. Use **Workbook 10** to find the cost saving opportunities described earlier:

1. Go to **Azure Monitor** and create a new empty workbook.

2. Next, click on **Advanced Editor** and select **Gallery Template**.

3. Copy and paste the code from `https://github.com/PacktPublishing/FinOps-Handbook-for-Microsoft-Azure/tree/main/Chapter-5` and select **Workbook 10**.

4. Click on **Apply** and then click on **Done Editing**:

Figure 5.21 – Workbook showing AKS clusters

5. Set **Workbook Auto Refresh** to **30 minutes**.

6. The workbook uses the following Azure Resource Graph query. You can certainly apply additional filters and conditions to suit your need:

```
resources
| where type == "microsoft.containerservice/
managedclusters"
| extend  AKSname=name,AKSID=id,location=location,AKSRG=
resourceGroup,Sku=tostring(sku.name),Tier=tostring(sku.
tier),
AgentPoolProfiles=properties.agentPoolProfiles
| mvexpand AgentPoolProfiles
| extend ProfileName = AgentPoolProfiles.name ,mode=tost
ring(AgentPoolProfiles.mode),AutoScaleEnabled = AgentPool
Profiles.enableAutoScaling , VMSize=tostring(AgentPool
Profiles.vmSize),minCount=tostring(AgentPoolProfiles.min
Count),maxCount=tostring(AgentPoolProfiles.maxCount) ,
nodeCount=tostring(AgentPoolProfiles.['count'])
| project AKSID,ProfileName,Sku,Tier,mode,AutoScale
Enabled,VMSize,nodeCount,minCount,maxCount,location,
AKSRG,AKSname
| join ( resources | mvexpand tags |
extend  AKSname=name,Tags=tags, resourceGroup
| extend tagName = tostring(bag_keys(tags)[0])
| extend tagValue = tostring(tags[tagName]) ) on AKSname
| project AKSID,ProfileName,Sku,Tier,mode,AutoScale
Enabled,VMSize,nodeCount,minCount,maxCount,tagName,
tagValue,location,AKSRG,AKSname
```

This concludes the top 10 cost optimization targets. Next, let's look at the trade-offs of cost and other aspects of the workload.

Trade-offs of cost versus security, performance, and reliability

When you architect the solution for Azure, consider trade-offs between cost and other aspects of design. To achieve the optimal cost, you must ask the business what is most important for the workload. Is it the lowest cost, no downtime, or high throughput? If you decide to go with the lowest cost, then the workload will not have multi-region high availability, nor can it have the best security controls against threats. And these are the risks you need to communicate to the business. As a rule of thumb, an optimal design doesn't equate to a low-cost design.

Cost versus security

Workload security is the most important aspect of hosting your code in the cloud. By nature, the public cloud is accessible over the internet, and securing your application requires the use of various services that then increase the cost of the solution.

For example, you are hosting your public-facing web applications to track an online payment and order system. Due to PCI requirements, you have to implement Azure Firewall to inspect the traffic, data encryption for SQL Server, just-in-time access for your IaaS VMs running batch jobs, and Web Application Firewall to prevent malicious attacks on the web applications. Apart from these services, you also have to use Microsoft Defender for Cloud to protect Azure PaaS services, data services, and networks. In this use case, each security service you consume costs additional money.

Cost versus performance

Higher performance requirements lead to increased costs. Processing incoming requests within a certain time requires you to provision faster, better, the latest, and sometimes specialty, hardware SKUs that increase the cost of the solution.

For example, let's say you are hosting a real-time data processing workload in Azure. The latency requirements between the user's request and the backend action are 2 seconds. Due to these business requirements, you have to use ultra disks that provide high throughput, high IOPS, and consistent low-latency disk storage with high-performance HB-series VMs optimized for memory bandwidth and HC-series VMs optimized for dense computation. This configuration certainly costs more than a regular VM.

Cost versus reliability

A workload's reliability refers to its ability to consistently serve customers and recover quickly in the event of a disaster. A business provides the required **Service Level Agreements (SLAs)**, **Recovery Time Objectives (RTOs)**, and **Recovery Point Objectives (RPOs)**.

For example, let's say a business has defined 99.99% SLAs for the mission-critical order processing system with 30 minutes of RTO and 30 minutes of RPO. To achieve the 99.99% SLA – which translates to 1 minute of downtime per week and 52 minutes of downtime per year. To achieve this level of availability, the solution must be hosted in two or more Azure regions. Apart from multi-region costs, you will also have to provision Traffic Manager, internal and external load balancers, autoscaling, monitoring, and alerting services. To achieve lower RTO and RPO, the database must be replicated and a backup has to be copied in multiple regions. All these extra services add costs to the original solution.

This concludes *Chapter 5, Hitting the Goals for Usage Optimization.*

Summary

In this chapter, we looked at the optimization phase of the FinOps lifecycle. The first step of the optimization journey starts with usage optimization. Inventory your current assets in Azure and find an opportunity to save money by avoiding costs or right-sizing the infrastructure to reduce wastage. We looked at how to define useful KPIs that help achieve the usage optimization goal. Then, we looked at the top 10 usage optimization targets and how to use the provided workbooks to bring visibility of these targets and take action accordingly. We concluded the chapter by pointing out that cost optimization is not without trade-offs. The cheapest solution will not give the best performance or protect against security threats.

Let's continue our journey to the second part of the optimize phase, rate optimization, in the next chapter.

Rate Optimization with Discounts and Reservations

Rate optimization refers to an activity where an organization works with the Microsoft sales team to secure better enterprise discount rates in exchange for a commitment to spend x amount over y period. An Enterprise Agreement, on the other hand, is a Volume license program where the customer is committed for a minimum of 3 years. Rate optimization can also be done via purchasing reservations. Azure Virtual Machines, Azure Storage Accounts, Azure SQL Database, Azure App Service, and many other services support reservations. Check out `https://github.com/PacktPublishing/ FinOps-Handbook-for-Microsoft-Azure/blob/main/Chapter-6/azure- services-list.md` for the complete list.

Purchasing reservations can be challenging, especially if you don't know what to reserve. Microsoft provides Azure Advisor, the Azure Cost Management Power BI app, and APIs to help you with recommended reservations you should purchase based on your past usage data. We will explore those in detail in this chapter. Once you have purchased a reservation, then you want to monitor the utilization. If the utilization is below 80%, you may want to exchange the reservation for something similar or return it. We will look at both the exchange and return process in detail.

In this chapter, we will learn about the following:

- Commitment-based discounts in Azure
- Identifying reservation opportunities for your workload
- Using the Azure Cost Management Power BI app
- Understanding Azure Advisor recommendations for reservations
- Reservation purchase and cadence
- Reservation details, renewal, savings, and chargeback reports
- Reservation exchange and cancellation

Let's get started!

Technical requirements

We will be using the following tools to accomplish the tasks in this chapter:

- The Azure Cost Management Power BI app, available at `https://learn.microsoft.com/en-us/azure/cost-management-billing/costs/analyze-cost-data-azure-cost-management-power-bi-template-app`.

- Azure Advisor, available at `https://ms.portal.azure.com/#view/Microsoft_Azure_Expert/AdvisorMenuBlade/~/score`.

- The Microsoft **Cost Management + Billing** tool, available at `https://portal.azure.com/#view/Microsoft_Azure_CostManagement/Menu/~/overview`. Alternatively, you can also find Cost Management + Billing by signing in to the Azure portal and, in the top center search bar, typing `Cost Management + Billing`.

When using these tools, sign in to the Azure portal using your organization ID. Please refer to *Chapter 1* for the minimum RBAC permissions required to use Azure **Cost Management + Billing**.

Commitment-based discounts in Azure

In a rate optimization exercise, commitment-based rate discounts are looked at first as they are foundational to your organization's journey in Azure, as well as the successful utilization of cloud services at scale. In this chapter, we will look at how you can get discounts in Azure via the Microsoft Enterprise Agreement and the Microsoft Azure Consumption Commitment program.

The Microsoft Enterprise Agreement

The Microsoft Enterprise Agreement is a volume licensing program that offers organizations with 500 or more devices the flexibility to purchase cloud services and software licenses under one agreement. The Enterprise Agreement is a 3-year agreement and allows you to forecast technology costs up to 3 years in advance. With Enterprise Enrollment and Server and Cloud Enrollment, you get the best pricing (discounts) for cloud-optimized licenses. You may get anything between 5% and 20% discounts for your Azure services with an Enterprise Agreement. When you order Enterprise Agreement services, you can pay everything up-front or pay in installments. Software Assurance is included with the Enterprise Agreement and provides the following benefits:

- Training vouchers
- E-learning
- Home use program
- 24x7 problem resolution support
- Windows virtual desktop rights

- License mobility through software assurance

- Azure Hybrid Benefit

To see the full list of Software Assurance benefits, review the Microsoft documentation at `https://www.microsoft.com/en-us/licensing/licensing-programs/software-assurance-default`

The Microsoft Azure Consumption Commitment (MACC)

The **Microsoft Azure Consumption Commitment** (**MACC**) is a contractual commitment to Azure spend over time. With MACC, there is no upfront payment required. In exchange for your commitment, you get a discounted rate for Azure and certain Marketplace purchases. You use MACC to improve the rates you pay for Azure services. Let's take an example of the MACC lifecycle and how the commitment is decremented.

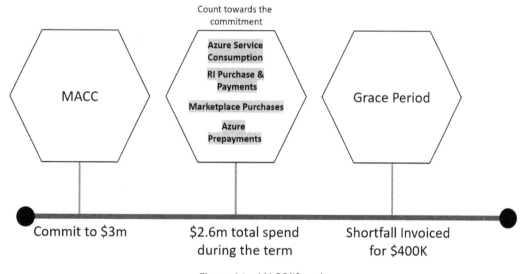

Figure 6.1 – MACC lifecycle

You have estimated that for the next 3 years, you are going to spend $3 million on Azure cloud services and third-party services such as Elastic Cloud and Datadog. You work with Microsoft to sign the MACC agreement that outlines all the commitments. Through this commitment, you get a percentage discount on your Azure cloud services. Now, anything you spend on Azure service consumption, reservation purchases and payments, eligible Marketplace purchases, and Azure prepayment will decrease the amount of your commitment. At the end of your 3-year commitment, if you have a shortfall, then that will be invoiced for prepayment. As you can see in this example, MACC is the best way to achieve rate optimization if you can commit to consumption. Now let's see how to find reservation opportunities for workloads.

Identifying reservation opportunities for your workload

How do you identify what reservations to purchase? This is the first question FinOps teams ask before discussing rate optimization. And it is important to be prepared to provide recommendations for reservation purchases to help the engineering team to come up with a business case. In the next chapter, we will look at how to write an effective business case for cost optimization.

To help with reservation purchases, Microsoft has provided two key resources. The first is the Azure Cost Management Power BI app. This app is a templated Power BI online application that offers the VM RI Coverage (Shared and Single) report, which shows regions, Azure service usage by pay-as-you-go price versus reservation price, and reservation recommendations. Apart from that, it also covers the following:

- Usage by subscriptions and resource groups
- Top five usage drivers
- Usage by services
- Azure Hybrid Benefit for Windows Server usage
- Reservation savings
- Reservation chargeback
- Reservation purchases
- Azure price sheet

The second resource to help with making a reservation purchase decision is Azure Advisor. Azure Advisor analyzes your configurations and usage telemetry and offers personalized, actionable recommendations to help you optimize your Azure resources for reliability, security, operational excellence, performance, and cost. As a FinOps team, we will focus on the cost pillar of Azure Advisor. Let's look at these resources next.

Using the Azure Cost Management (ACM) Power BI app

The ACM Power BI app's VM RI Coverage report queries all VMs in your subscriptions and provides matching recommendations for replacing pay-as-you-go VMs with reservations. The recommendation is calculated based on evaluating the last 30 days of usage.

Before you install the ACM Power BI app, you will need the following:

- A Power BI Pro license
- The Enterprise Administrator (read-only) role to connect to the data

To install the ACM Power BI app, please follow the Microsoft documentation located at `https://learn.microsoft.com/en-us/azure/cost-management-billing/costs/analyze-cost-data-azure-cost-management-power-bi-template-app`. Once the Power BI app is installed, you can share it with the rest of the users in your organization.

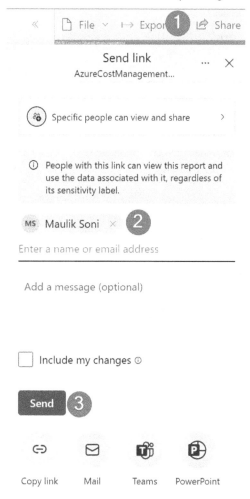

Figure 6.2 – Share the ACM Power BI app with your team

First, click on the **Share** button in the report toolbar. Then provide the name of the person in your organization that you want to share the app with and click **Send**.

You can also keep your data up to date by setting up **Scheduled refresh**. Go to the **ACM Dataset** settings and turn on **Scheduled refresh**.

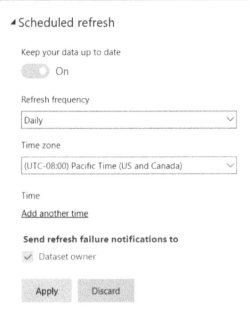

Figure 6.3 – Schedule data refresh for the ACM Power BI app

Now let's look at the report and find out what reservations to purchase.

Scenario 1 – you are purchasing VM reservations for the first time

You currently do not have any reservations across your Azure enrollment. This is the first time you are purchasing VM reservations. In this case, the VM RI Coverage report will show only on-demand usage data.

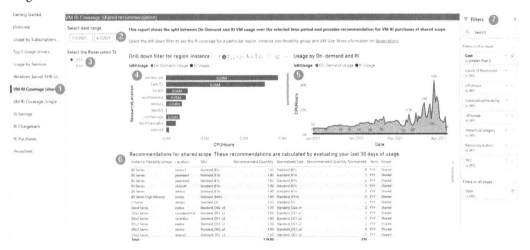

Figure 6.4 – VM RI Coverage recommendation

Let's understand how to use this report:

1. The left side of the report is the navigation to switch between different reports bundled in the ACM app.

2. The date range filter will filter the results based on the selected date.

3. The reservation term can be 1 year or 3 years. You can switch between the two and see how much you can save. Selecting a 3-year reservation will give you greater savings.

4. **Drill down filter for region, instance size, and VM selection**: Select the region that has the most on-demand costs. For example, in the preceding report, we have most of our resources in West Europe. This means when we are ready for a reservation purchase, we will be purchasing it in the West Europe region.

5. The **Usage by On-demand and RI** graph tells you how much of your spend is on-demand pricing versus reservation pricing. Since, in this case, I do not have any prior reservations, I see the light-colored graph showing on-demand costs.

6. And finally, we have the **Recommendations for shared scope** table. In the Shared scope, the reservation discounts are applied to matching resources in all eligible subscriptions. The opposite of shared scope is single-subscription scope, where a reservation discount is applied to the matching resources in the selected subscription. It is always recommended to purchase reservations for shared scope for better flexibility.

7. Additional filters can be used to filter recommendations by SKU or meter subcategory.

Now, you are tasked with making the first purchase. You will first select the region in the region drill down. That will filter the recommendation table for that specific region.

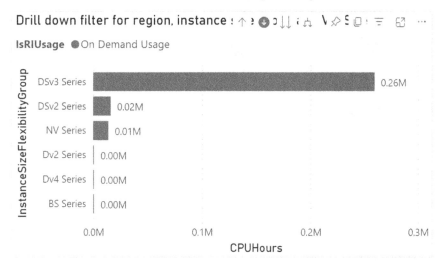

Figure 6.5 – Region drill down for reservation recommendations

As you can see, in the West Europe region, we are spending about $200,000 on DSv3 series VMs. Click on **DSv3 Series** in the drill down graph again. We will see the following table updated to show DSv3 instance family reservations.

Recommendations for shared scope. These recommendations are calculated by evaluating your last 30 days of usage.

Instance Flexibility Group	Location	SKU	Recommended Quantity	Normalized Size	Recommended Quantity Normalized	Term	Scope
DSv3 Series	westeurope	Standard_D2s_v3	1.00	Standard_D2s_v3	1	P1Y	Shared
DSv3 Series	westeurope	Standard_D4s_v3	3.00	Standard_D2s_v3	6	P1Y	Shared
DSv3 Series	westeurope	Standard_D8s_v3	3.00	Standard_D2s_v3	12	P1Y	Shared
Total			**7.00**		**19**		

Figure 6.6 – Virtual machine reservation recommendations

Based on the recommendations, we will purchase 19 normalized quantity of Standard_D2s_v3 VM SKU to cover the usage by the instance family of D2s_V3, D4s_v3, and D8s_v3. We will look at the reservation purchase experience in the *Reservation purchases* section later in this chapter.

Scenario 2 – you have existing reservations but want to purchase a new one for another Region and VM SKU

You have purchased reservations in the past but they do not cover the current workload. In this case, the VM RI Coverage report will show a breakdown of on-demand usage and reservation-covered usage data. Your goal is to close the gap between the on-demand price you pay for the workload and insufficient reservations that have been purchased in the past.

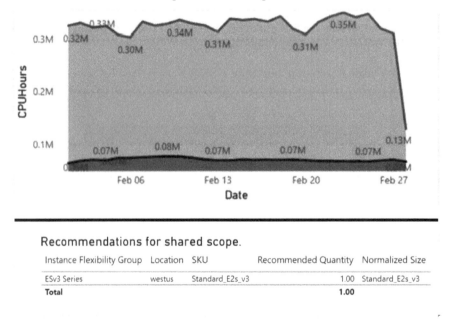

Recommendations for shared scope.

Instance Flexibility Group	Location	SKU	Recommended Quantity	Normalized Size
ESv3 Series	westus	Standard_E2s_v3	1.00	Standard_E2s_v3
Total			**1.00**	

Figure 6.7 – Virtual machine reservation coverage for esv3-Series Virtual machine

Note the preceding **Usage by On-demand and RI** graph. There is a light area and a dark area. The light area of the graph represents the on-demand price you pay for the VMs. The dark area is the portion of VMs covered under your current reservation purchase. In other words, you have a lot of opportunities to save money by purchasing more reservations. As you purchase more reservations, the dark area of the graph will expand the coverage and nearly meet the peaks of all your VM pricing.

In this case, **Recommendations for shared scope** tells us to purchase one Standard E2s_v3 reservation for a 1-year term. We will look at the reservation purchase experience in the *Reservation purchases* section later in this chapter.

The Azure Cost Management Power BI app is not the only tool that helps with reservation decisions. Azure Advisor also plays a critical role. Let's look at Azure Advisor recommendations next.

Understanding Azure Advisor recommendations for reservations

Azure Advisor is an excellent companion to view your reservation recommendations in an easy-to-understand format and provides excellent savings details. To get to Azure Advisor, in the Azure portal's top search bar, search for `Advisor`. Once Azure Advisor opens, select **Cost** under the **Recommendations** tab on the left.

Figure 6.8 – Azure Advisor showing reservation recommendations

In Azure Advisor, you can select a reserved instance configuration to suit your needs. In this example, we have selected 3 years for the reservation term and 30 days for the look-back period to examine our usage.

Let's look at the first recommendation – **Buy virtual machine reserved instances to save money over pay-as-you-go costs**. Click on the link in your recommendation and it will take you to the **Recommendation details** screen as shown.

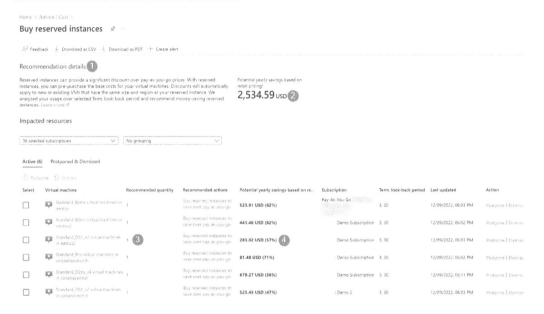

Figure 6.9 – Reservation recommendations details

Let's look at the main elements of the **Buy reserved instances** screen:

1. **Recommendation details** will provide a brief description of the reservation purchase recommendation.

2. Azure Advisor calculates the overall potential yearly savings with these reservations based on your resource SKU and retail or pay-as-you-go price.

3. Recommended quantities are calculated again based on your past usage data. In our case, we selected a look-back period of 30 days.

4. Each recommendation will show the potential annual savings based on the VM SKU and reservation discount.

Azure Advisor gives you the unique ability to also purchase a selected reservation right from these recommendations by clicking on the **Recommended actions** link. Once you click on the link, it will take you to the purchase experience shown here:

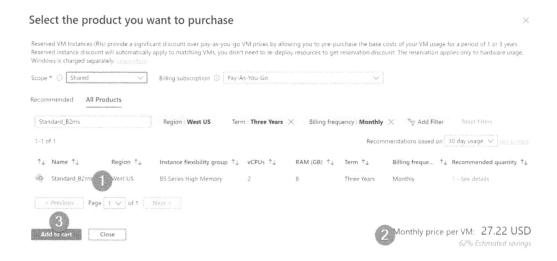

Figure 6.10 – Add reservation to cart

Let's examine the main elements of reservation purchase page:

1. First, make sure the reservation details are correct and, in fact, that's the reservation you want to purchase.

2. The monthly price per VM will show the actual price after reservation discounts along with the percentage discount applied.

3. When you are ready, click on **Add to cart**.

4. And finally, click on **Review + buy** to purchase the reservation.

Congratulations…! You have purchased your first Azure reservation and started saving money. In the next section, we will look at the purchase experience in a bit more detail.

Reservation purchase and cadence

All reservations are applied on an hourly basis (except for Databricks). When you are purchasing reservations, it should be based on your consistent usage. You can determine which reservations to purchase by examining the usage data and leveraging Azure Advisor and the ACM Power BI app, or even by using reservation APIs.

> **Note**
>
> If you want to learn more about how to get a list of recommendations using a REST API, you should check out the Microsoft documentation available at `https://learn.microsoft.com/en-us/rest/api/consumption/reservation-recommendations/list`

Reservation purchase is certainly not a one-time task. As a guiding principle, when you are starting your journey, start small. Purchase a first reservation for a small quantity of VMs to get yourself comfortable with the lifecycle of reservations. As time goes by, you will be purchasing reservations more frequently to cover newer workloads.

Reservations are charged to the payment method attached to the billing subscription you select during purchase. If your organization has a monetary commitment, then reservation charges are deducted from that. For up-front payments, your account will be charged immediately. If you have selected the monthly payment option, then you will see a monthly charge on your invoice.

To purchase a reservation, do the following:

1. Go to the Azure portal and in the top search bar, type `Reservations`.

2. On the **Reservations** screen, click on **Add**.

3. As of writing this book, there are 27 services that support reservation pricing and Microsoft keeps expanding and including new services.

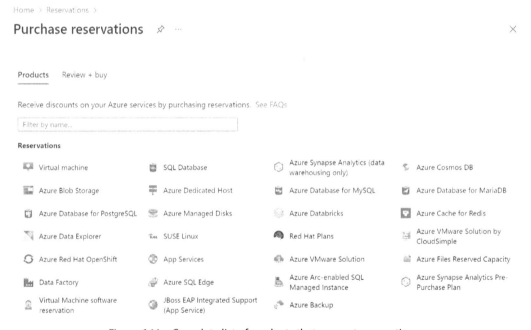

Figure 6.11 – Complete list of products that support reservations

4. Click on **Virtual machine** from the **Products** list to go to the details screen.

5. **Reservation details** has two tabs at the top. By default, the **Recommended** tab is selected, which shows the SKU for the VMs that you should reserve. You can select **All Products** to see all the VM SKUs.

Select the product you want to purchase   ✕

Reserved VM Instances (RIs) provide a significant discount over pay-as-you-go VM prices by allowing you to pre-purchase the base costs of your VM usage for a period of 1 or 3 years. Reserved instance discount will automatically apply to matching VMs, you don't need to re-deploy resources to get reservation discount. The reservation applies only to hardware usage. Windows is charged separately. Learn More

Scope * ⓘ [Shared   ∨]   Billing subscription ⓘ [Budget Demo   ∨]

Recommended   All Products

[Filter by name, region, or instance ...]   Region : **East US 2** ✕   Term : **Three Years** ✕   Billing frequency : **Monthly** ✕   Reset filters

                              ⁺▽ Add Filter

1-3 of 3   Recommendations based on [30 day usage ∨] Download as CSV | Learn more   (●) Optimize for instance flexibility (preview)

↑↓	Name ↑↓	Re... ↑↓	Instance flexibility group ↑↓	vC... ↑↓	R... ↑↓	Te... ↑↓	Billing freque... ↑↓	Recommended quantity ↑↓
🖥	Standard_DS1_v2	East US 2	DSv2 Series	1	3.5	Three Y...	Monthly	2 - See details
🖥	Standard_D2s_v3	East US 2	DSv3 Series	2	8	Three Y...	Monthly	2 - See details
🖥	Standard_D2_v2	East US 2	Dv2 Series	2	7	Three Y...	Monthly	1 - See details

Figure 6.12 – Selecting D-Series VMs for reservations

6. Note the billing subscription will be the subscription in which you will see the reservation purchase charges. Up-front or monthly, both charges will show up on the invoice of your selected billing subscription.

7. Now, let's reserve **Standard_DS1_v2 VM** with a quantity of 2. Select the line and it will calculate the monthly price per VM and the estimated savings as a percentage. In this case, our VM will cost around $17.69 per month and we are getting 57% savings over the pay-as-you-go price.

1-3 of 3   Recommendations based on [30 day usage ∨] Download as CSV | Learn more   (●) Optimize for instance flexibility (preview)

↑↓	Name ↑↓	Re... ↑↓	Instance flexibility group ↑↓	vC... ↑↓	R... ↑↓	Te... ↑↓	Billing freque... ↑↓	Recommended quantity ↑↓
🖥	Standard_DS1_v2	East US 2	DSv2 Series	1	3.5	Three Y...	Monthly	2 - See details
🖥	Standard_D2s_v3	East US 2	DSv3 Series	2	8	Three Y...	Monthly	2 - See details
🖥	Standard_D2_v2	East US 2	Dv2 Series	2	7	Three Y...	Monthly	1 - See details

< Previous   Page [1 ∨] of 1   Next >   Not seeing what you want? Browse all products.

[Add to cart]   [Close]   Monthly price per VM: 17.69 USD
                                                                57% Estimated savings

Figure 6.13 – Reservation screen showing the monthly price and estimated savings

8. You can further click on the **Recommended quantity** tab to see even more details about the reservation you are purchasing. You will be able to see the total monthly cost (of all VM quantities) and the total reservation cost.

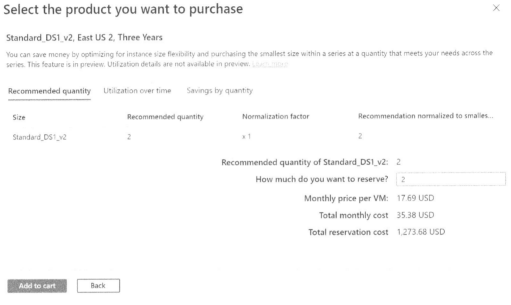

Figure 6.14 – Adding reservation to cart

9. Click on **Add to cart** to proceed.

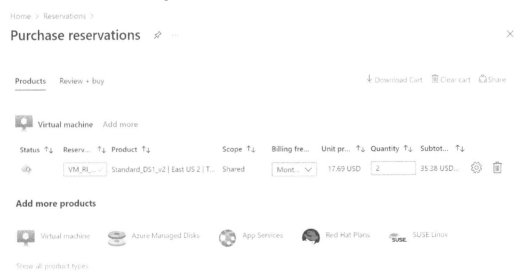

Figure 6.15 – Edit reservation details

10. At this point, you can still edit the reservation name, change the billing frequency from monthly to up-front, or adjust the quantity. If you no longer want to proceed with this reservation, you can click on the *Delete* icon and it will remove this item from the cart.

11. You can click on **Review + buy** and finally click on the **Buy Now** button to finalize the purchase. You will see the following screen, which shows the progress of the purchase:

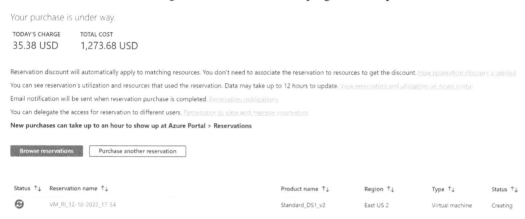

Figure 6.16 – Reservation purchase being processed

12. Once the reservation is successful, you will see this reservation listed on your **Reservation** screen.

In this example, we looked at reserving a VM. In a similar fashion, you can purchase Azure Blob Storage, Data Factory, or App Service, or any of the other 27 services that offer reservations. Check out the Azure Storage reservation experience at `https://github.com/PacktPublishing/FinOps-Handbook-for-Microsoft-Azure/blob/main/Chapter-6/storage-reservation.md` and the Azure App Service reservation experience at `https://github.com/PacktPublishing/FinOps-Handbook-for-Microsoft-Azure/blob/main/Chapter-6/appservices-v3-reservation.md`.

Purchase cadence

Your workload in Azure is constantly changing. The engineering team is enhancing existing applications and services and building new ones and deploying them to Azure. In this constantly changing cloud environment, reservation purchase is certainly not a one-time activity. In fact, it's a continuous process to make sure you pay less for on-demand pricing and purchase reservations and get rate discounts to cover at least 80% of your entire workload in Azure.

There are essentially two models for purchase cadence. First, FinOps teams decide to purchase reservations on a quarterly, semi-annually, or annual basis. Now, this fixed time-based approach has the advantage of assessing the workload over a period of time and then purchasing reservations. This approach has the trade-off that, to purchase reservations, you have to wait for the cadence and some of your workload will not get the benefit of the rate discount early in the lifecycle. To overcome this effect, more mature organizations evaluate the need for reservations and purchase them as and when needed. This gives better coverage for new workloads to apply the discounted rate from the beginning.

Next, let's look at how to view reservations that you or your team have purchased and find out how are they being utilized.

Reservation details, renewal, savings, and chargeback report

After you have purchased the reservation, you want to see all the reservation details, its renewal settings, and utilization reports. In this section, we will learn how to accomplish these tasks.

Reservation details

To view the reservation details, take these steps:

1. Go to **Reservations** in the Azure portal.

2. The **Reservations** screen will list all the reservations that you have purchased.

Figure 6.17 – List of reservations

This list shows important information about your reservations. The **Status** column provides the status of the reservation. In the preceding example, it says it expires on November 3, 2022. We can also see that the automatic **Renewal** status is Off. This means once this reservation expires, the VM it covers will be charged at the on-demand price.

3. Click on the **7 Day Utilization percent** link. It provides the following details about the usage of the reservation.

Figure 6.18 – Reservation utilization details

This reservation has only been utilized 13.04% in the past 7 days, which is a serious issue. This is a form of *vacancy* in FinOps terms, and as a FinOps practitioner, you should strive to ensure that all reservations reach their full utilization potential of 100%. Unused reservations are a waste, and it is your job to make sure this does not happen.

If you have noticed that utilization is lower than desired, it is important to review usage data to make sure that all applicable usage is receiving its reservation discount. If you find that reservations exceed usage, you can either execute a partial or full reservation exchange or resize deployments to match the reservation. In addition, investigate how the reservation discount is applied, and consider expanding the scope if it is limited to a single subscription or resource group.

Auto-renewal

At the end of the reservation term, the billing discount expires. Your resources continue to run normally and are then billed at the pay-as-you-go rate. You can certainly enable the auto-renewal of reservations. To enable auto-renewal, go to your reservation and click on **Renewal** on the left, under **Settings**. Check **Automatically renew this reservation** and then click the **Save** icon at the top.

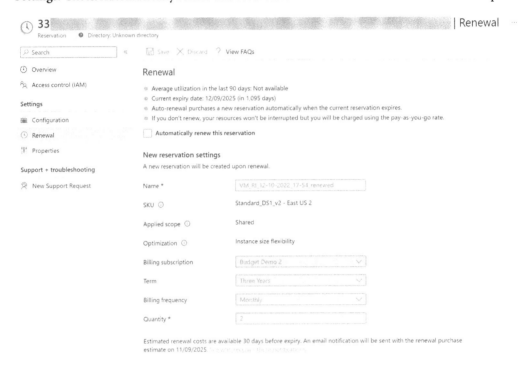

Figure 6.19 – Reservation renewal

When the reservation expires, it can be renewed with a new term, billing frequency, and quantity. It is recommended to enable reservation auto-renewal to avoid paying the regular pay-as-you-go price for your Azure resources.

Reservation savings and chargeback report

Now that you have purchased reservations, the finance department is asking for the report that shows how much money we are saving and how we are going to charge it back. The ACM Power BI app provides both reports.

Go to the ACM Power BI app and then select **RI Savings reports** from the left menu. For the date range, we have selected Q1 (Jan 1 – March 31) to see the quarterly savings.

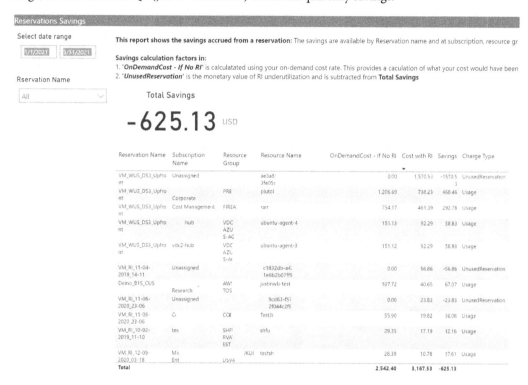

Figure 6.20 – Reservation savings report in the Azure Cost Management Power BI app

The report displays the cost of not having any Reserved Instance (**If No RI**) as well as the cost with Reserved Instances. The savings for having Reserved Instances can be found by subtracting the *If No RI cost* from the *RI savings cost*. This report also shows unused reservations. You want to dig deep because, essentially, it means there is a vacancy in your reservation and you have a loss if you are not utilizing the reservation at 100%.

The RI chargeback report will help you understand the extent of the RI benefits applied to which subscription and resource group.

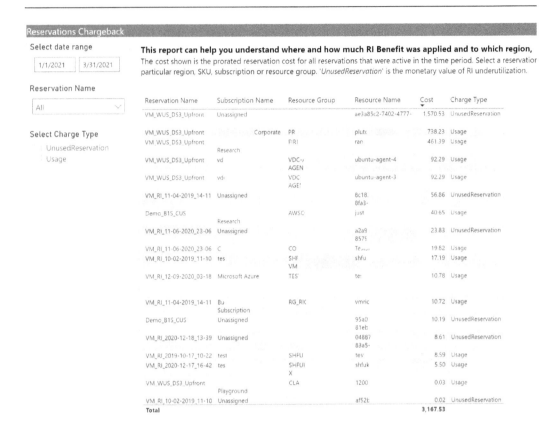

Figure 6.21 – Reservations chargeback report

You can use this report to charge back a subscription that belongs to a given department. Let's now look at how to exchange or cancel a reservation.

Reservation exchange and cancellation

As we looked at in the previous section, if you find that reservations you have purchased are not being utilized as anticipated, then you can decide to exchange the reservation for a like-to-like new reservation or decide to cancel it.

Exchange reservations

Azure reservations provide flexibility to meet your changing needs. Reservations are interchangeable for the same type of reservation. During an exchange, you can change the SKU, Region, scope, and term. For example, you can exchange a reservation for SQL Managed Instance in East US with SQL Elastic Pool in West US. Note, however, that you cannot exchange a SQL reservation with a VM reservation.

To exchange a reservation, follow these steps:

1.   Go to the **Reservations** screen and click on **Exchange**:

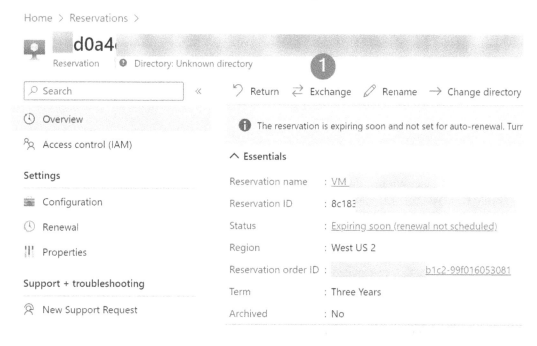

Figure 6.22 – Reservations screen

2.   Select the reservation you want to exchange:

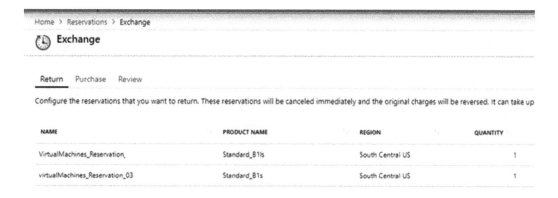

Figure 6.23 – Exchange existing reservation

3.   Next, select a new reservation you want to purchase. Make sure the purchase total is more than the return total:

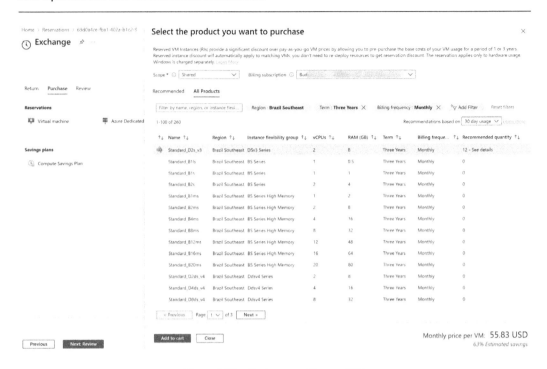

Figure 6.24 – Exchange with new SKU

4.   Review and complete the purchase:

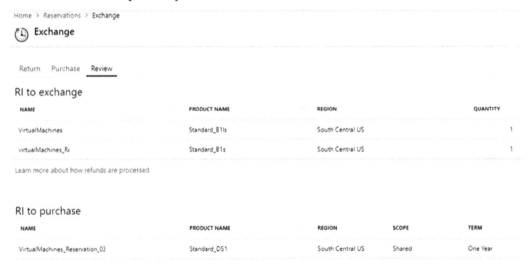

Figure 6.25 – Finalize the exchange

To refund a reservation, go to **Reservation Details** and select **Refund**. The sum total of all cancelled reservation commitments in your billing scope cannot exceed USD 50,000 in a 12-month rolling window.

Cancel (return) a reservation

To cancel (return) a reservation, follow these steps:

1. Go to **Reservations** and select the specific reservation. Then click on the **Return** button.

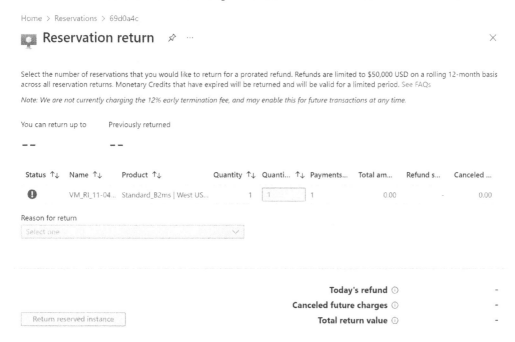

Figure 6.26 – Reservation return screen

2. Click **Return reserved instance** to get the refund. Please note that refunds are limited to $50,000 on a rolling 12-month basis.

As you can see, reservations in Azure provide an excellent way to save money by optimizing the rate you pay for cloud services in exchange for usage commitment. They are also flexible enough to exchange and cancel. Let's summarize what we learned in this chapter.

Summary

In this chapter, we looked at rate optimization using the Enterprise Agreement, MACC, and reservations. The Enterprise Agreement and MACC provide organizations with a better-discounted rate for Azure services. These commitment-based agreements are the first step for FinOps teams to work toward if they are not in place. Even if an organization does have an EA or MACC, when it's time to renew, finance, procurement, and engineering teams can negotiate better rates by committing to sustained use of Azure services.

We looked at reservations in detail. What to reserve is the biggest question FinOps teams run into and Microsoft provides Azure Advisor, the ACM Power BI app, and APIs to get recommendations based on your usage in the last 7, 30, or 60 days. Once you purchase reservations, you should check the utilization of the reservations and monitor them for any changes that might be needed. You can either return reservations if no longer needed or you can exchange them for something similar.

In the next chapter, we'll look at some real-world optimization strategies to get better rates.

7
Leveraging Optimization Strategies

Cost optimization efforts should not be focused on one strategy or feature. Instead, it should be evaluated holistically. Azure provides various ways to save money by removing waste, right-sizing, purchasing reservations, savings plans, and highly discounted Spot VMs.

Let's first explore highly discounted Spot VMs. They are excess Azure capacity available to you with deep discounts without any commitment. The only caveat is that compute can be taken away within 30 seconds of notice so the workload has to be architected in such a way that can handle the eviction. Before you purchase the Spot VMs, you need to decide on the region and SKU since the prices vary based on those. Azure does provide an excellent view of the past 3 months of Spot VM prices and the rate at which eviction was executed. Use this key information to decide the region and SKU selection.

On the other hand, Spot Priority Mix is the best of both worlds. It's a mix of Spot and regular VMs in VM Scale Sets. It provides consistent compute capacity in addition to additional Spot VMs.

Azure savings plan is a new and exciting offering that is an alternative to reservations. We will look in detail at how reservations and savings plans compare and when to use what.

In this chapter, we will learn about the following:

- Introducing Azure Spot market
- Spot VM caveats
- Pricing history and eviction rate details
- Architecting the workload to handle eviction
- Spot Priority Mix
- Discounting strategies with savings plans
- Writing a business case for cost optimization

Let's get started!

Technical requirements

We will be using the following tools to accomplish the tasks in this chapter:

- Azure Advisor available at `https://ms.portal.azure.com/#view/Microsoft_Azure_Expert/AdvisorMenuBlade/~/score`

- Microsoft **Cost Management + Billing** tool available at `https://portal.azure.com/#view/Microsoft_Azure_CostManagement/Menu/~/overview`. Alternatively, you can also find **Cost Management + Billing** by signing in to the Azure portal and in the top-center search bar, type `Cost Management + Billing`.

When using these tools, sign in to the Azure Portal using your organization ID. Please refer to *Chapter 1* for the minimum RBAC permissions required to use Azure **Cost Management + Billing**.

Introducing Azure Spot market

Azure Spot market offers unused Azure compute capacity (VMs) at a steep discount of up to 90% compared to pay-as-you-go pricing. Any time Azure needs the capacity to support its pay-as-you-go customers, it will evict the Spot VM with 30 seconds' notice. Due to that fact, Spot VMs are best used with a workload that can tolerate the interruption, such as Databricks clusters, Azure Kubernetes clusters, image rendering, or batch processing jobs. There are no SLAs for Spot VMs and the excess capacity can vary between regions, VM SKUs, and time of the day.

Estimating Spot VM discounts

Before you purchase Spot VMs, let's estimate how much discount we can get using the Azure pricing calculator:

1. Open the browser.

2. Navigate to the Azure pricing calculator site at `https://azure.microsoft.com/en-us/pricing/calculator`.

3. Select the options **CentOS**, **East US 2**, **United States – Dollars**, **Month**, and **Dsv3-series**, as shown in the following screenshot:

Figure 7.1 – Azure pricing calculator for Spot VMs

4. In the following list, note the **Pay as you go** versus **Spot** pricing discount details:

Instance	vCPU(s)	RAM	Temporary storage	Pay as you go	Spot ▲
D2s v3	2	8 Gib	16 Gib	$70.08/month	$30.84/month 56% savings
D4s v3	4	16 Gib	32 Gib	$140.16/month	$61.67/month 56% savings
D8s v3	8	32 Gib	64 Gib	$280.32/month	$123.34/month 56% savings
D16s v3	16	64 Gib	128 Gib	$560.64/month	$246.68/month 56% savings
D32s v3	32	128 Gib	256 Gib	$1,121.28/month	$493.36/month 56% savings
D48s v3	48	192 Gib	384 Gib	$1,681.92/month	$740.04/month 56% savings
D64s v3	64	256 Gib	512 Gib	$2,242.56/month	$986.73/month 56% savings

Figure 7.2 – Spot VM price list

5. For a D8s v3 VM with 8 CPUs and 32 Gib RAM, the **Pay as you go** price is **$280.32/month**, while the **Spot** instance price will be **$123.34/month**, which is a saving of 56%.

Use the Azure calculator to estimate your savings with Spot instances. Azure also provides pricing details during Spot VM creation, including personalized recommendations as well as a few other important details. Let's look at that next.

Spot VM and VM Scale Sets

Azure Spot discounts are available for individual VMs or VM Scale Sets. The difference between creating a single VM with a Spot discount versus VM Scale Sets depends on the workload as well as the ability to scale in and out instances. For demonstration, we will use VM Scale Sets and will show the Azure platform's **Try & Restore** feature for Spot VMs.

To create a VM Scale set with a Spot discount, follow these steps:

1. Open the Azure Portal.

2. Search for `Virtual Machine Scale Set` in the top search bar.

3. Click **Create**.

4. In the **Basics** tab, check the box next to **Run with Azure Spot discount**.

Figure 7.3 – Create a virtual machine scale set

5. Once you check the box, the **Azure Spot configuration** details will be displayed.

Figure 7.4 – Spot VM configuration details

6. Click on **Configure** to adjust the Spot instance details. It will take you to the **Spot details** tab, as shown in the following screenshot:

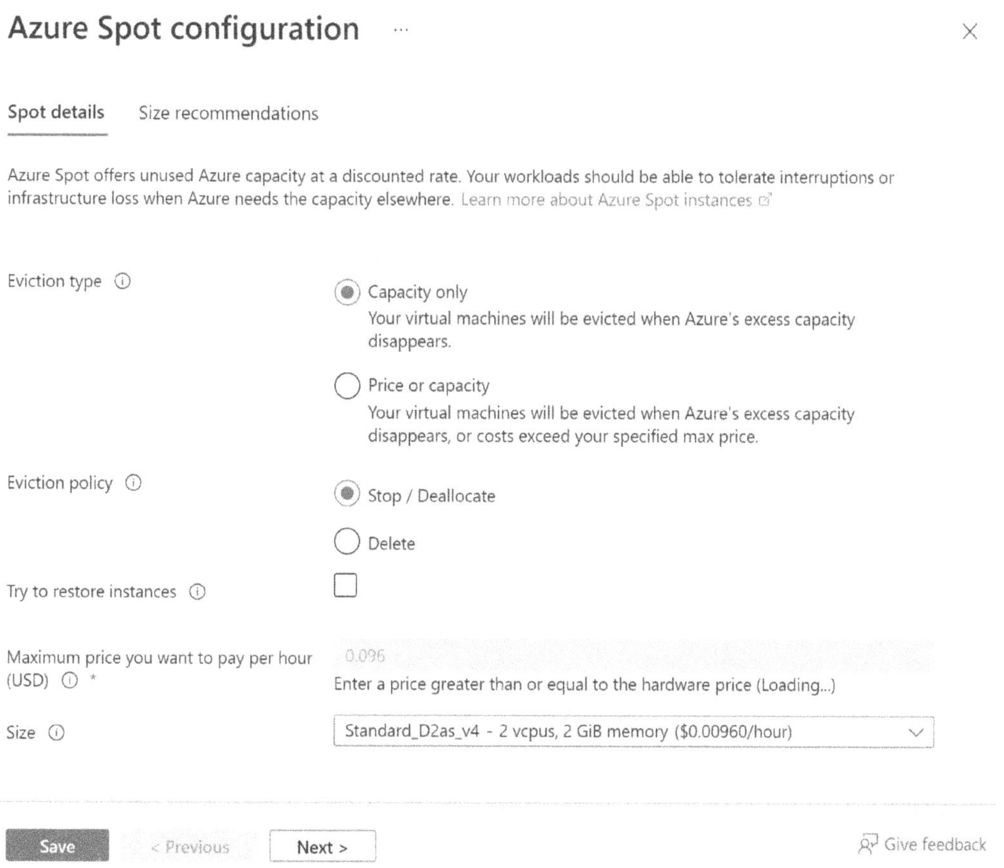

Azure Spot configuration ...   ✕

Spot details   Size recommendations

Azure Spot offers unused Azure capacity at a discounted rate. Your workloads should be able to tolerate interruptions or infrastructure loss when Azure needs the capacity elsewhere. Learn more about Azure Spot instances ⬀

Eviction type ⓘ	◉ Capacity only
	Your virtual machines will be evicted when Azure's excess capacity disappears.
	○ Price or capacity
	Your virtual machines will be evicted when Azure's excess capacity disappears, or costs exceed your specified max price.
Eviction policy ⓘ	◉ Stop / Deallocate
	○ Delete
Try to restore instances ⓘ	☐
Maximum price you want to pay per hour (USD) ⓘ *	0.096
	Enter a price greater than or equal to the hardware price (Loading...)
Size ⓘ	Standard_D2as_v4 - 2 vcpus, 2 GiB memory ($0.00960/hour) ⌄

Save   < Previous   Next >   ᗉ Give feedback

Figure 7.5 – Eviction type and policy selection

Azure Spot VM configuration has the following options:

- **Eviction type** – Choose **Capacity only**. This means that when Azure needs the capacity, it will evict your VM and allocate it to other pay-as-you-go customers.

- **Eviction policy** – Choose what happens to VMs that are evicted. Upon eviction, the VMs can either be stopped and deallocated if you plan to re-deploy, or deleted.

- **Try to restore instances** – This platform-level feature will use AI to automatically try to restore evicted Azure Spot VM instances inside a scale set to maintain the target instance count.

7. To change the size of the VM, click on **Size recommendations**:

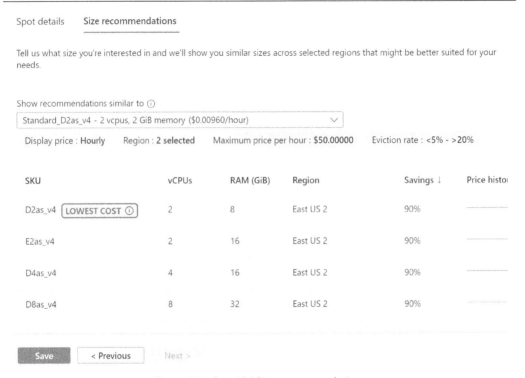

Figure 7.6 – Spot VM Size recommendations

8. In the **Size recommendations** screen, you can see the SKU with the **LOWEST COST** label. This is the SKU that has the lowest price point. When selecting the right SKU, take into consideration that the price history over 90 days is stable and the eviction rate is < 5%.

9. Click **Save** and continue the steps to create the VM scale set.

Once the VM scale set is created, you can go to the **Instance details** page and confirm that Spot VMs are being used.

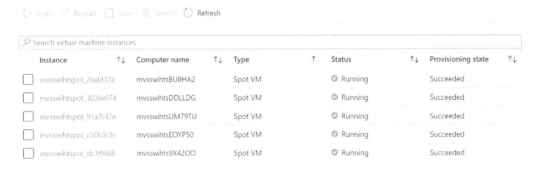

Figure 7.7 – VM scale set running with Spot VM instances

You can also simulate the eviction and test how well your application handles it. Use the following Azure PowerShell command:

```
Set-AzVM -ResourceGroupName "mySpotRG" -Name "mySpotVM"
-SimulateEviction
```

To use the REST API endpoint to simulate eviction, refer to `https://learn.microsoft.com/en-us/rest/api/compute/virtual-machines/simulate-eviction?tabs=HTTP`.

While Spot VMs offer excellent discounts on the price, there are limitations that you need to keep in mind and design workloads that can handle the caveats. Let's look at that in detail next.

Spot VM caveats

Spot VMs are not without trade-offs. Evaluate whether your workload can benefit from the price discount without impacting its availability, reliability, and security aspects. Let's understand the eviction policy and limitations of spot VMs.

Eviction type and policy

What is eviction? Spot VMs do not have any **service-level agreement** (**SLA**) and they can lose the compute anytime with 30 seconds' notice. This loss is called **eviction**. Eviction is driven by supply and demand in a specific region. When a certain VM SKU demand goes too high, the platform starts evaluating the Spot VMs to accommodate new pay-as-you-go VMs. Spot VMs have two configurations for eviction: **Eviction type** and **Eviction policy**.

Eviction type defines the condition of the eviction. There are two types – **Capacity only** and **Price or capacity**. When Azure's excess compute capacity disappears, it triggers the **Capacity only** eviction in which Spot VM prices are not fixed and can change. The second type – the **Price or capacity** eviction – happens when Azure changes the price of the Spot VM and it is beyond your maximum allowed price.

Eviction policy determines what happens to the VM when Azure triggers the eviction of your Spot VM. There are two policy options. The first is **Stop / Deallocate**, which means that when eviction happens, Azure will stop the VM and deallocate the non-static IP but will keep the disk. This policy is suitable for a workload that can wait for the next available capacity in the same region and VM type. The second option – the **Delete** VM policy – will delete the VM and data disks. This is most suitable for workloads that can be relocated to different regions and VM sizes.

Limitations

Azure Spot VM does not support B-series (Burstable) or any promotional version of VM SKU, for example, NC, and H promo series. In terms of region restrictions, Spot VMs can be deployed in any region except China 21vianet.

Apart from these, at its core, there is no high availability guarantee with Spot VMs. That's the major limitation you may want to consider before migrating your workloads to Spot VMs.

Before you decide to use Azure Spot VMs for your workload, you may want to look at the pricing history for your Azure region and the rate at which Spot VMS were evicted.

Pricing history and eviction rate details

Azure Spot instance prices are variable, based on region and SKU. To provide customers with some visibility, Azure provides the price history of the last 90 days. Use the pricing history details to select the right VM SKU and region where prices are stable and cheap.

To view the Spot VM pricing history, follow these steps:

1. Go to the Azure Portal.

2. Create a new VM with **Run with Azure Spot discount** checked:

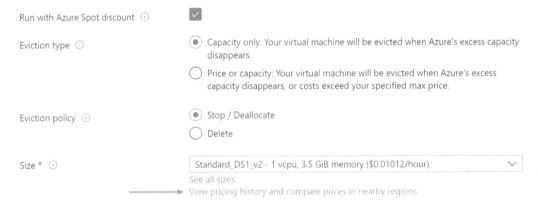

Figure 7.8 – Run VM with Spot discount

3. Click on the **View pricing history** link and it will open up the pricing details page:

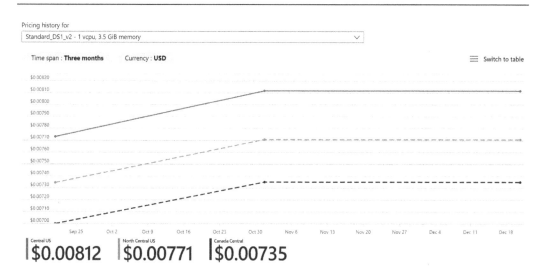

Figure 7.9 – Spot VM pricing history for the selected VM SKU

Here, you can see a price graph for the selected VM for three nearby regions. You can also switch the view to a table.

In the example, as you can see, the price increased at the end of October 2022 from $0.00773 to $0.00812 in Central US.

4. Next, to view the Spot VM eviction rate, click on the **See all sizes** link:

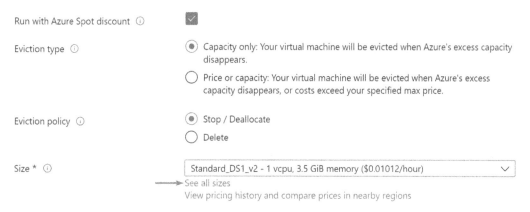

Figure 7.10 – View additional Spot VM sizes

The **Eviction rate** column displays the percentage of VMs that were evacuated due to capacity being reclaimed in the past 28 days.

VM Size ↑↓	Type ↑↓	vCPUs ↑↓	RAM (GiB) ↑↓	Data ... ↑↓	Max IOPS ↑↓	Te... ↑↓	Premium disk ↑↓	Eviction rate ↑↓	Cost/hour ↑↓
∨ Most used by Azure users ↗			The most used sizes by users in Azure						
DS1_v2 ↗ ⓘ	General pur...	1 (Customize)	3.5	4	3200	7	Supported	0-5%	$0.01012 (87%)
D2s_v3 ↗ ⓘ	General pur...	2	8	4	3200	16	Supported	0-5%	$0.01717 (85%)
D2as_v4 ↗ ⓘ	General pur...	2	8	4	3200	16	Supported	0-5%	$0.01680 (85%)
DS2_v2 ↗ ⓘ	General pur...	2	7	8	6400	14	Supported	5-10%	$0.02224 (85%)
D4s_v3 ↗ ⓘ	General pur...	4	16	8	6400	32	Supported	0-5%	$0.03333 (86%)

Figure 7.11 – Eviction rate history

When deciding the VM SKU, factor in the eviction rate. The lower the rate, the higher the possibility that your VM will not be interrupted by eviction.

Even if we select the lowest eviction rate, our application must handle the eviction gracefully. Let's look at how to architect applications with best practices to handle evictions.

Architecting the workload to handle eviction

Eviction is the process whereby Azure takes back your Spot VM when Azure's excess capacity disappears. Workloads running on Spot VMs must accept the possibility of eviction and should have the following characteristics:

- Low importance in the organizations and no up-time constraints
- Run processes that are idempotent, stateless, and short
- Run processes that can stop/restart without losing any data

Examples of interruptible workloads are batch processing applications, data analytics, and workloads that create a continuous integration/continuous deployment agent for a non-production environment.

Let's look at a real-world scenario to understand the architecture better. The image resizer application is a stateless batch processing job that runs when an image in a storage account is uploaded in the **source** container by the frontend application. It uses the open source **FFmpeg** utility to resize the image and uploads it back to the **resized** container in the same storage account. To deploy the application on a Spot VM instance, we have come up with the following architecture:

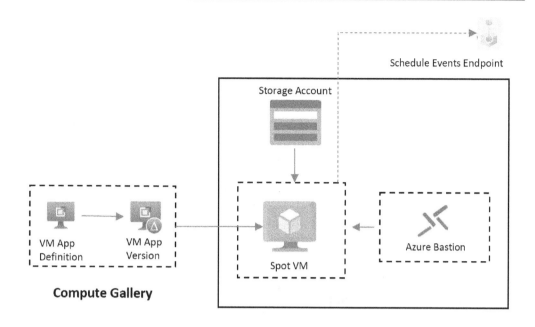

Figure 7.12 – Architecture pattern for an interruptible workload

Here are the key features of our application architecture:

- **Use Compute Gallery to package and distribute the application**. For Spot VMs, compute is very valuable. We want to minimize the time it takes between eviction and fully running the application by pre-installing all required software and its dependencies.

- **Use the schedule events endpoint to continuously monitor for eviction**. Azure sends out signals to VMs when they're going to be affected by infrastructure maintenance. It sends out the pre-empt signal to all VMs at a minimum of 30 seconds before they're evicted. In our architecture, we created a small service that runs every second and queries the non-routable IP address 169.254.169.254 for the eviction signal. Once it receives the signal, it gracefully shut down our image resizer service and stops taking new work.

- **Test the architecture by simulating the eviction event**. You can either use the REST endpoint or AZ CLI to simulate the scenario and test how well your application handles it.

- **Designing an idempotent workload is key**. In our scenario, the resize command is executed for a single image file. Running the command for each image file gives us two benefits over looping over a collection of images. First, it takes very little time to resize a single image. Second, if the VM is evicted, then there is no complex state to be stored since we are operating on a per-file basis.

But what if you want to have a pool of VMs with a mix of Spot and regular VMs to support the continuity of your workload? Let's look at the Spot Priority Mix feature next.

Spot Priority Mix

Spot Priority Mix (public preview) is a feature that allows you to mix and match regular VMs with Spot VMs in a VM scale set and gives a cost advantage. Spot Priority Mix only works with the flexible orchestration mode of VM Scale Sets. Using this feature, you can reduce your compute cost by adding discounted Spot VMs to the scale set, and at the same time, you can still maintain minimum compute capacity by allocating minimum uninterruptible VMs. This will guarantee that all your VMs won't be taken away at the same time as part of the eviction process. Your leftover regular VMs will be continuing to serve requests while new VMs are being provisioned.

A custom percentage distribution can be configured across Spot and regular VMs. Scale-out and scale-in operations are orchestrated automatically to achieve the desired distribution by selecting the appropriate number of VMs to be created or deleted.

Let's look at how to create a VM scale set with Spot Priority Mix:

1. Log in to the Azure Portal.
2. In the search bar, search for and select **Virtual Machine Scale Sets**.
3. Click on the **Create on Virtual Machine Scale Set** page.
4. On the **Basics** tab, select **Flexible** as the **Orchestration** mode.
5. On the **Scaling** tab, check the box for **Scale with VMs and Spot VMs**:

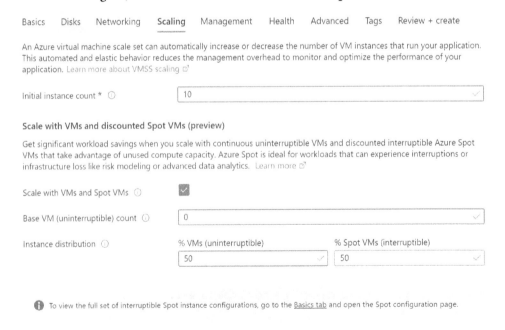

Figure 7.13 – VM scale set with Spot and regular VMs

6. Click **Review + create**.

In the preceding example, we are creating 10 initial VMs for our data analytics workload. Out of those 10 VMs, 5 will be regular or uninterruptable VMs and 5 will be Spot VMs. As auto-scaling occurs, the flexible distribution method will keep track of the distribution percentage and will create the next VM, balancing the distribution.

Azure provides Spot VM size recommendations that allow you to select alternative VM sizes. Azure makes recommendations based on your region, price, and eviction rate. You can further refine the recommendations based on the size, type, generation, and disk (premium or ephemeral OS disk). Refer to the Microsoft documentation at `https://learn.microsoft.com/en-us/azure/virtual-machine-scale-sets/spot-vm-size-recommendation` to learn more about Spot VM sizes.

Spot VMs are excellent for workloads that can be interrupted. But there are times when you want to look for guaranteed compute for workloads and to be able to save money. This is when Microsoft savings plans come into the picture. They are different than reservations and offer unique scenarios to save money on globally distributed workloads. Let's look at savings plans next.

Discounting strategies with savings plans

Azure savings plan for compute became **generally available (GA)** in Oct 2022 and provide savings of between 11% and 65% over pay-as-you-go pricing. Essentially, Azure savings plan for compute (or for short, savings plan) is a fixed hourly commitment for you for 1- to 3-year terms. You can pay upfront or monthly at no extra cost.

Let's understand the savings plan using an example provided by Microsoft:

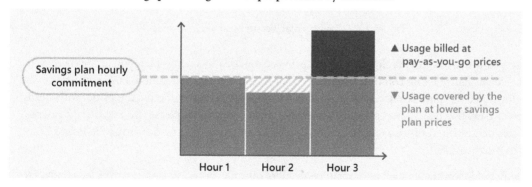

Figure 7.14 – How the savings plan discount is applied

When you commit $5 per hour to a savings plan for 1 year, the following scenarios can happen:

- If your hourly usage is $5, then all your usage is covered by discounted savings plan pricing
- If your hourly usage is above $5, then over-usage will be charged at a pay-as-you-go rate
- If your hourly usage is below $5, then it's an unused savings plan

> **Note**
>
> Azure savings plan for compute includes eligible resources for five Azure services such as App Service, Azure Container Instances, Azure Dedicated Host, Azure Functions, and Azure Virtual Machines. For detailed information, please review the Microsoft documentation at `https://learn.microsoft.com/en-us/azure/cost-management-billing/savings-plan/savings-plan-compute-overview`.

Important to note is that once you purchase the savings plan, you cannot cancel it. Exercise caution when choosing the right hourly commitment amount during purchase.

Savings plan versus reserved instances

Let's look at how savings plans are different from reserved instances. Let's look at the brief comparison provided by Microsoft:

	Compute Savings Plan	Reserved Instance
Savings compared to pay-as-you-go	Save up to 66%	Save up to 72%
Commitment	Fixed hourly dollar amount	Specific virtual machine type in a specific region
Savings apply	Across participating compute services globally on hourly basis	Directly to the specified compute service in a selected region
Term	1 or 3 year	1 or 3 year
Payment	Upfront or monthly	Upfront or monthly
Cancellation	No cancellation allowed	Up to $50,000 USD

Figure 7.15 – Savings plan versus reserved instances

Reserved instances provide slightly higher savings compared to savings plan but, at the same time, provide less flexibility for compute service, VM types, and regions. You want to buy reservations first for the stable compute resources and then look for dynamic resources (across the globe) with steady spending and purchase savings plans. To learn more about savings plan, please visit `https://azure.microsoft.com/en-us/pricing/offers/savings-plan-compute/#benefits-and-features`.

Highly stable workloads such as domain controllers running 24x7 and with no expected changes to the VM SKU or region can benefit from reservations. While dynamic workloads (such as a help desk case management system deployed across the globe) can run on different VM SKUs and accommodate changes, regions can benefit from a savings plan.

Purchasing a savings plan

To purchase a savings plan, follow these steps:

1. Go to the Azure Portal.

2. Search for `Savings Plans` in the top search bar.

3. Click on **Purchase Now**.

4. Next, provide the following details:

   I. Select your **Billing subscription** preference.

   II. Select **Shared across the subscription's billing scope**.

   III. For maximum savings, select a 3-year term.

   IV. For **Hourly commitment in USD**, use the Azure-provided personalized recommendations based on recent usage.

   V. **Cost summary** will display the monthly as well as the total cost of the savings plan.

   VI. Click **Next: Review + buy** to purchase the savings plan.

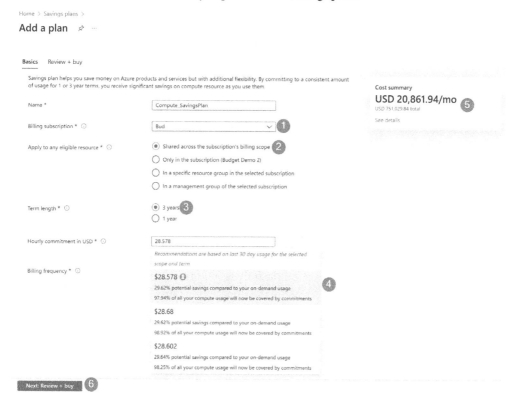

Figure 7.16 – Purchase new savings plan

After you have purchased the savings plan, you can track the utilization of the plan on the **Overview** page of the savings plan or in the **Azure Cost Management** tool using the **Amortized cost** view.

For a limited time, you may be eligible to trade in reservations for savings plans; so far, the exchange is equal to or greater in total value. Check with your Microsoft sales representative for more information.

Now that we have seen what savings plans are, the finance team needs to decide and approve the purchase. In the next section, we will look at how to write a business case.

Writing a business case for cost optimization

A business case is a justification of why the business needs to make an investment in a cost optimization project or purchase a cost optimization third-party product. You can make a case by listing financial benefits and/or non-financial benefits, such as improved visibility in cloud spending or optimizing cloud operations. The business case highlights the cost and benefits of implementing the project. Sometimes, it also refers to a cost/benefit analysis report.

In this example, we will focus on writing a business case for the cost optimization of an existing big data analytics workload and how it can financially benefit the organization to optimize the current cloud spend over 1-year and 3-year terms.

Business case: Orion business analytics platform cost optimization

Executive summary: The "Orion" business analytics workload currently costs $12 million/year for its infrastructure hosting in the Microsoft Azure cloud. The price we pay is a pay-as-you-go price without any discounts, and it's the highest price any customer can pay. There is a significant savings opportunity (>50%) with this workload by exploring reservations, savings plans and possibly leveraging Spot Priority Mix VMs.

Problem: "Orion" is a business analytics workload running in Azure and utilizes VMs, storage accounts, Databricks clusters, AKS clusters, SQL Server Managed Instance, App Services, and Function Apps.

The current run rate of the total cost for this workload is $1 million per month. For this workload, we are paying a street price, or in other words, a pay-as-you-go price.

Opportunity: The "Orion" workload can significantly benefit from rate optimization strategies. Microsoft Azure provides three ways for which we can improve the rate we pay for our infrastructure in Azure. Reservations can provide savings of up to 72%, a savings plan can provide savings of up to 66%, and Spot instances can provide savings of up to 90% compared to the current pay-as-you-go price.

The following is the current breakdown of the "Orion" workload's cost categorization:

$1,000,000 / Month

$200,000 Unpredictable Usage
$500,000 Dynamic Resources Consistent Spend
$300,000 Fixed Compute Resources

Pay-as-you-go Pricing

Figure 7.17 – Pay-as-you-go pricing for the workload

Benefits: The following is the proposed breakdown of reservation discounts and savings plans. We are anticipating to purchase $150,000 of Reservation of fixed compute resources and $200,000 of Savings Plans for dynamic resources.

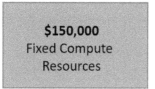

$150,000 Fixed Compute Resources	**$200,000** Dynamic Compute Resources

Reservation Discounts   Savings Plan Discounts

Figure 7.18 – Reservation and savings plan discounts for the workload

Overall savings: The business case proposes overall savings of $450,000 by purchasing a combination of Reservation and Savings Plans.

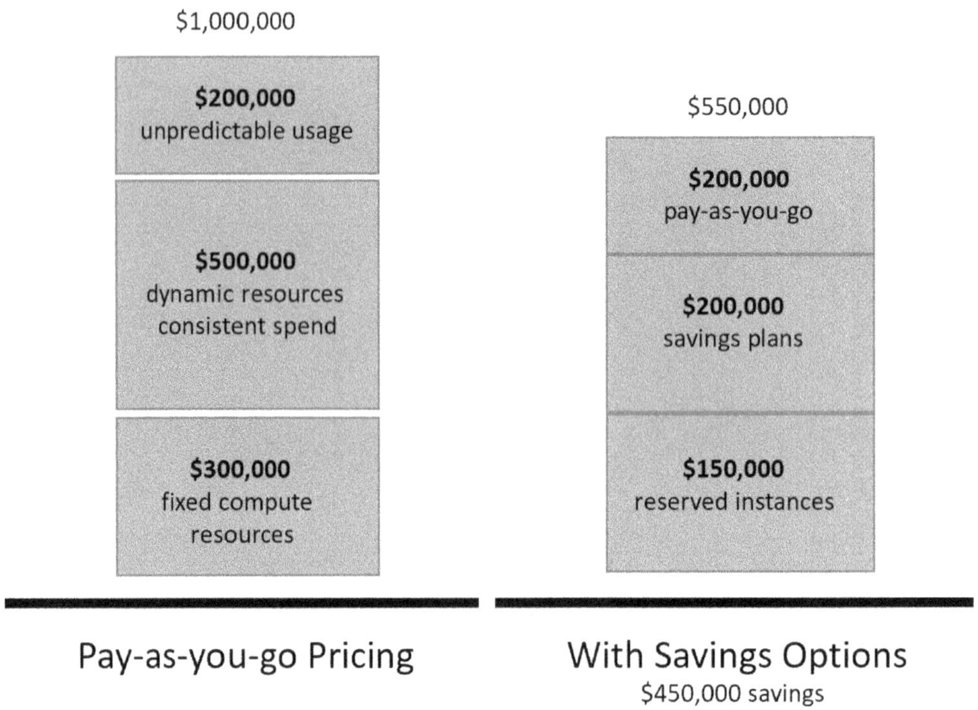

$1,000,000

$200,000 unpredictable usage
$500,000 dynamic resources consistent spend
$300,000 fixed compute resources

$550,000

$200,000 pay-as-you-go
$200,000 savings plans
$150,000 reserved instances

Pay-as-you-go Pricing   ## With Savings Options

$450,000 savings

Figure 7.19 – Pay-as-you-go pricing versus savings options comparison for the workload

Assumptions: We have the following two assumptions:

- "Orion" will continue to spend on cloud infrastructure at the baseline rate identified at $1 million/month for the next 3 years

- "Orion" will expand to other business units resulting in an increase in cloud infrastructure spending by 30% of the baseline spend

Constraints: These are the constraints:

- The cost optimization target is only for the "Orion" business analytics workload

- Reservations are commitments for 1 to 3 years

- Savings plans cannot be canceled once purchased

- Spot instances can be evicted by the Azure platform if excess capacity disappears

- Some workloads may need to be re-architected to run on Spot instances

Risks: Let's look at the risks:

- If the demand for the "Orion" system decreases, then we end up having higher reservations than needed. This is called **vacancy**, and it will have a negative financial impact on the organization.

- Reservations can be canceled or exchanged with a cancellation charge of $50K or exchanged for the same or a higher value reservation, which also resets the reservation term.

Implementation plan and milestones: Here is how this will be implemented:

- A new V-team will be formed to execute this project, which includes representation from finance, procurement, FinOps, engineering, and product team members

- Milestone 1: Purchase 3-year reservations for 50% of VMs within 5 days of business case approval

- Milestone 2: Purchase 3-year reservations for 80% of SQL Managed Instance databases within 10 days of business case approval

- Milestone 3: Purchase a savings plan for global compute resources with a $5-an-hour commitment for 3 years within 30 days of business case approval

- Milestone 4: Configure Databricks instances, Azure Kubernetes Service clusters, and VM Scale Sets to use Priority Mix with Spot VMs within 30 days of business case approval

Recommendation: Based on the extensive analysis of the "Orion" workload and available rate optimization opportunity in Azure, it is recommended to purchase a mix of reservations, savings plan, and Spot instances to achieve the maximum savings of $450,000/year.

Once the business case has been approved by the finance team, the engineering team can start planning its implementation. The cost optimization user stories are written and prioritized. And finally, the necessary purchases are made.

With this, we are at the end of *Chapter 7*. Let's summarize what we have learned.

Summary

In this chapter, we looked at various alternatives to usage optimization and reservations. Spot VMs are an excellent way to get deep compute discounts on excess Azure capacity without any commitment. On the other hand, savings plans are a brand new Azure feature and work with the same as reservations. There are distinct advantages to savings plans over reservation. Architecting workloads for Spot VMs can be different, so we looked at the ideal architecture and key characteristics that a workload should exhibit to take advantage of Spot VMs. And lastly, we looked at how to write a cost optimization business case and present it to the finance team. As a FinOps practitioner, you will be frequently doing this and hopefully, the template we have provided will be a good head start in this direction.

Congratulations...! You have completed the **Optimize** phase of the FinOps life cycle. What could be more relevant than a case study? Let's look at how Peopledrift Inc. used rate optimization to achieve its savings goals.

8

Case Study - Realize Savings and Apply Optimizations

In this chapter, we will review the healthcare company case study. This is the second case study out of three case studies in this book. Each case study reinforces the learning from the previous section. In this case study, we will continue to witness Peopledrift Healthcare's journey to enhance the FinOps team's capabilities and achieve the goals of the Optimize phase.

Now, it's been more than a year since Peopledrift Healthcare strategically adopted the Microsoft Azure cloud and migrated the majority of their on-prem workload and transformed the entire organization according to a cloud-first model. The Microsoft Azure cloud is used to host **virtual machines** (**VMs**) and Active Directory domain servers, store immense amounts of data, and run serverless workloads. The consistent and high usage of cloud services now challenges the FinOps team to look for opportunities and strategies to save money. To overcome all these hurdles, the FinOps team needs to optimize various rates and usage areas.

In this case study, we will cover the following:

- Finance wants innovative ways to control and reduce the cost of the cloud due to consistently high usage

- What KPIs has Peopledrift designed to measure the progress of savings?

- How was the usage optimization program designed and executed?

- How was the rate optimization program designed and executed?

- Is an Azure **savings plan** (**SP**) worth considering?

Case study – Peopledrift Inc., a healthcare company

We are a healthcare company specializing in providing supporting IT services to hospitals and doctors' offices all around the world. Our **Online Appointment Scheduling** (**OAS**) platform is highly rated by the healthcare industry, and we are the leader in this space. Three years ago, we acquired a small transport and logistics company now powered by our DeliverNow online platform, which delivers vaccines at speed, accuracy, and scale to hospitals, pharmacy stores, and doctors' offices.

In the past year, we strategically moved out of a data center and selected Microsoft Azure as a strategic cloud provider for all our workloads. We migrated a total of 150 internal and customer-facing applications and 90 relational and non-relational, extremely large databases. The estimated capacity numbers provided by engineering grew significantly as we expanded our business to new horizons. This has caused us to continuously provision large infrastructure in Azure.

Challenges

After migrating our entire data center workload to Azure as an IaaS service, we immediately realized that the total cost of running IaaS services is way more than initially thought. The service usage and rate we pay for cloud services need to be investigated and business cases need to be prepared to optimize the cost. This is when our CEO, CTO, and CFO asked the newly formed FinOps team to step in and provide systematic programs to evaluate, initiate, and execute cost optimization efforts to reduce the current cost of cloud services.

Objectives

Peopledrift Healthcare has identified the following key objectives for optimization:

- The FinOps team will collaboratively define **Objectives and Key Results** (**OKRs**) and KPIs that help with rate and usage optimization

- We will utilize Azure Advisor recommendations and track cost score improvements over time

- We will define and run a systematic usage optimization program to achieve the OKRs and KPIs defined by the organization

- We will identify rate optimization opportunities for Microsoft Azure and execute programs to realize savings

The solution

We adopted a FinOps framework some time ago and established the foundational processes. As our utilization of Azure grew exponentially, so has our FinOps team. We trained two internal resources and hired one FinOps expert to strengthen the team and work on more complex and high-priority initiatives.

After witnessing tremendous growth for the past 2 years, the rate of adoption of our platform has finally started to slow down, which reflects the overall economic condition across the business. Most of our customers have either started scaling down the number of platform licenses or slowing down in purchasing new licenses. This slowdown in business mixed with an unfavorable economic outlook has created financial challenges for Peopledrift. During the last financial strategy meeting, the CEO and CFO laid out aggressive but realistic plans to achieve savings across the organization.

The plan is outlined as 0-50-40-10 in terms of cloud resource cost. Zero wastage of cloud resources. 50% of the cost savings will be attained through the reservations purchased for fixed workloads, 40% of the savings will be achieved through savings plans for dynamic workloads, and only 10% of the total cloud cost will be paid as regular PAYG pricing.

The FinOps team has created four distinct programs and formed a **virtual team** (**v-team**) for each to work toward the savings goals defined by the leadership team. The v-teams comprise financial representatives, procurement representatives, product team members, and engineering team members.

To measure the real-time progress of the initiatives, the FinOps team defined the following individual KPIs and created a combined dashboard to reduce the waste and realize savings by purchasing reservations and savings plans:

- Zero waste – Dev/test resources without auto-shutdown during the weekend

- Zero waste – VMs without Azure Hybrid Benefit

- Zero waste – SQL Server without Azure Hybrid Benefit

- Zero waste – Number of unattached disks

- Zero waste - Number of unattached public IPs

- Zero waste – Number of AKS clusters without auto-scaling

- 50% savings – Moving from PAYG costs to reservations

- 50% savings – Fixed workload savings report (amortized)

- 40% SP – PAYG cost of dynamic resources versus a dynamic workload savings plan

- 40% SP – Dynamic workload savings report (amortized)

- 10% Pay-as-you-go cost – Total cloud cost versus pay-as-you-go price report

The FinOps team developed a report, published it to Power BI online, and shared the report with the v-team and leadership team to track progress. It is important to note here that the KPIs were defined before the execution of the 0-50-40-10 initiative. Without KPIs, the organization would have no way to track progress.

For the execution of this cost-saving initiative, the FinOps team has created a project plan with milestones to achieve in 45 days. Four separate v-teams were formed and each one was assigned one initiative based on its expertise.

The zero-waste v-team

The objective of this v-team is to remove all the waste in the current dev/test and production environment hosted in Azure. As we saw the rapid adoption of Microsoft Azure, unprecedented new resources were created and updated. During this life cycle, we found various resources that were simply provisioned but never used. We discovered numerous Azure firewalls and virtual private network gateways, ExpressRoute circuits, VMs, databases, and AKS clusters that had been set up but were no longer in use for various reasons. The zero-waste team conducted an in-depth examination of all resources in terms of their utilization, by using custom Azure Monitor workbooks to pinpoint the top 10 waste targets and employing Kusto Query Language. The second part of cutting back on waste was rightsizing and shutting down VMs and databases, through which the v-team were able to save approximately $45,000 on a monthly basis, by keeping track of and eliminating waste.

The RI savings v-team

The objective of the Reserved instance savings v-team is to assess, identify, purchase, and manage Azure reservations. Azure reservations provide an excellent way to save money on a fixed or consistent workload. The RI savings v-team created the following KPIs to track the progress of savings along the way:

- Number of resources for which reservations can be purchased
- Number of reservations recommended by business unit
- Number of reservations purchased month over month
- Reservation utilization versus vacancy
- 1-year cost versus 3-year reservation savings comparison
- PAYG versus reservation savings month over month

First, the v-team utilized various Azure-provided tools to identify reservation opportunities. The team used the **Azure Cost Management** (**ACM**) Power BI app's RI recommendation report (shared and single-scope). Along with that, it also utilized Azure Advisor recommendations for each business unit and created visibility into what reservations procurement should purchase.

The team knows that the reservation purchases are not a one-time job. Initially, they went with a quarterly reservation assessment and purchase schedule. This allowed enough time for everyone on the team to learn about and understand the nuances of reservations and amortized cost reporting. The more advanced method of purchasing reservations will be metric-driven, which they will implement early next year.

To view the amortized costs, the team used the ACM Power BI connector and utilized the RI usage summary and RI usage details tables. The RI savings report from the ACM Power BI app was also leveraged from time to time for reporting the savings goals.

The SP v-team

The SP v-team's objective is to identify any additional or complementary savings that could be achieved by purchasing a Microsoft Azure SP. In essence, SPs are special discounts you get when you commit to a certain amount of $ per hour of usage. Please note that you cannot cancel or exchange SPs, so be wise when committing to the $/h amount.

The SP v-team first identified the dynamic workload. In other words, for a workload that is spread across the globe, its utilization pattern is dynamic. The team found out the customer service application was the best candidate for an SP due to its servers being distributed across the globe and it utilizing VMs, App Services, Azure Functions, and Container Instances. The team used the Azure recommendation for the hourly committed $ amount provided in the **Azure Savings Plan** blade in the Azure portal.

The SP v-team also used its PAYG versus SP savings KPI to keep track of the SP's potential and the actual savings report.

The 10% PAYG v-team

The objective of this v-team is simple – measure the total Microsoft Azure cost that Peopledrift pays at the end of the month. Then, break down the total cost into three buckets – the PAYG price, the reservations price, and the SP price. At the end of the month, the PAYG price should be only 10% of the total cost. To understand better, let's look at the KPI that the FinOps team developed. In this graph, we have stacked the total cost of reservations and SP discounts. Combining both, only 10% of the cost should be paid as the PAYG price while the rest of the cost should be covered by the rate optimization plans.

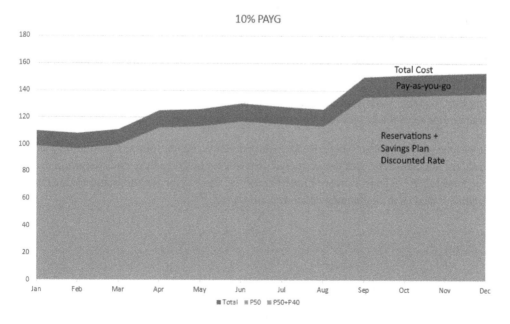

Figure 8.1 – Only 10% of the total cost should be at a PAYG rate

This wonderful report shows month-over-month costs comparing the rate. As you can see, **Reservations + Savings Plan** covers 90% of the rate at a discount while we are only paying the remaining 10% at the **Pay-as-You-Go** rate. By monitoring this report, the FinOps team can continuously fine-tune the reservations, and either purchase a new one or cancel/exchange existing ones based on its utilization data.

Alternatively, we also used the Azure Cost Management Power BI app's usage-by-demand and RI report to see how much of a rate is covered by reservations, although this report does not consider the SP rate.

At the end of 45 days, all the v-teams successfully completed the execution of their assigned optimization initiatives, resulting in 50% savings in VM costs and 40% savings in App Services costs. Overall, it was a huge success!

Benefits

The benefits procured by practicing the Optimize phase of FinOps practice are as follows:

- The FinOps team is now able to handle usage and rate optimization efforts and help the business achieve savings goals in the currently difficult economic climate.

- The collaborative practice of FinOps has brought the finance, procurement, product, and engineering teams together to achieve key business objectives.

- The systematic execution of the Well-Architected Framework assessment, Azure Advisor recommendation reviews, the top 10 usage optimization targets, and remediation allowed Peopledrift to achieve 20% savings by removing waste and rightsizing its resources.

- The FinOps team was able to achieve 60% savings by purchasing reservations for 3 years. The team also recommended purchasing SPs at a $27 commitment per hour for 3 years, resulting in 55% savings for global dynamic resources in Azure.

- The finance team was able to see the cost optimization per department, business unit, and application using the savings report created by the FinOps team.

- The FinOps team built a monthly PAYG versus reservation savings report showing how reservations are utilized and how they affect payment rates.

Our FinOps journey does not stop here. Keep reading the next chapter for tips for building a culture of FinOps, cost allocation for containers, metric-driven cost optimization, and designing metrics for unit economics. You'll learn how we implemented the Operate phase and matured our FinOps practice, which saved us in challenging economic times.

Summary

In this case study, we examined Peopledrift Healthcare's second phase of transformation, which was Optimize. After a year of effort to migrate the entire data center to Azure, fueled by business growth, the cost of Azure services has become a source of expenditure panic for the company's leadership. Add economic uncertainty to that and it's a perfect recipe for fear among leadership, but FinOps is meant to deal with this scenario with crystal-clear guidance on how to systematically approach it.

Peopledrift essentially formed and focused its v-teams around four distinct areas, starting with an initiative called zero waste in Azure. This v-team focused on rightsizing, removing unused resources, and auto-scaling services. The second v-team focused on reservation opportunities – what to reserve, how much, and when. The third v-team was laser-focused on a brand-new Azure feature called SPs. The final v-team kept track of new utilization and whether it went beyond 10% PAYG pricing, then altered the reservations, and the SP team took action.

In the next chapter, let's look at building a culture of FinOps. We will investigate key characteristics that make FinOps practices successful and the most rewarding investment for organizations.

Part 3: Operate

This part provides a comprehensive introduction for building effective FinOps teams and fostering a collaborative work culture. Moreover, you will explore the primary difficulty in cost allocation for containers and examine its resolution. You will also gain insights into the automated assessment of cloud infrastructure, the utilization of metrics to optimize expenses, and the development of unit economics metrics to correlate cloud expenditure with business outcomes. To solidify your understanding, a case study is presented at the end.

This part contains the following chapters:

- *Chapter 9, Building a FinOps Culture*
- *Chapter 10, Allocating Cost for Containers*
- *Chapter 11, Metric-Driven Cost Optimization*
- *Chapter 12, Developing Metrics for Unit Economics*
- *Chapter 13, Case Study – Implementing Metric-Driven Cost Optimization and Unit Economics*

9
Building a FinOps Culture

Without a process for actions, applied goals are never achieved. As we take the journey into the *Operate* phase, building a collaborative culture across various business boundaries is essential for the continued success of FinOps practice.

The culture of FinOps is collaboration. It is the FinOps team's responsibility to bring the stakeholders together and prepare a business plan that clearly articulates savings opportunities. Building such a team requires management buy-in. Management must be fully convinced that establishing a **Center of Excellence** (**CoE**) for cloud cost management is the way to control and optimize cloud costs in the modern world. Engineers nowadays can create cloud infrastructure with automation tools such as Terraform, without realizing the total cost of ownership of that infrastructure. And it is only the FinOps team who can glue together the finance, business, product, and engineering teams to spend money on what matters the most. We will also review what motivates the engineering team to take cost-saving actions and when to use the carrot approach versus the stick approach. We will also cover cost automation opportunities and review Azure native cost management tools and briefly look at third-party cost optimization solutions.

In this chapter, we will learn about the following:

- Establishing a CoE for cloud cost management
- Motivating engineering teams to take action
- Automated tag inheritance, governance, and compliance
- Automated VM shutdown and startup
- Automated budget actions
- Third-party FinOps tools

Let's get started!

Technical requirements

We will be using the following tools to accomplish the tasks in this chapter:

- The Microsoft **Cost Management + Billing** tool, available at `https://portal.azure.com/#view/Microsoft_Azure_CostManagement/Menu/~/overview`. Alternatively, you can also find Cost Management + Billing by signing in to the Azure portal and, in the top-center search bar, typing `Cost Management + Billing`.

When using these tools, sign in to the Azure portal using your organization ID. Please refer to *Chapter 1* for the minimum RBAC permissions required to use Azure **Cost Management + Billing**.

Establishing a CoE for cloud cost management

A CoE for cloud cost management becomes a necessity when an organization spends a good amount of money on the cloud. During economically uncertain times, when cost-cutting measures are implemented across the organization, the cloud is the first target.

To appropriately handle financial challenges, a structured and centralized organization is the desire of leadership. The skills needed to build the FinOps teams are still very scarce, and experienced analysts are hard to find.

It is recommended to at least hire an experienced person who has implemented cost savings measures to jump-start the journey. The cloud cost management field is huge and requires specific cloud service provider knowledge. Once the FinOps lead is hired, you can immediately plan an upskilling program for one or two other employees who may be interested in this field. Having a team of at least three people is recommended.

Once the FinOps team takes shape, start with small initiatives. Initially, everyone will be new, and processes will not have been fully vetted. Keep in mind that the FinOps team and its work are highly collaborative. It is relationship building that will take some time and, along the way, earning trust is very important. The FinOps team must have empathy toward every person they interact with and communicate the message that they are the enabler of cloud cost optimization, which is in the interest of the organization and everyone else.

As the CoE gains experience in executing small business cases and building bridges between various teams, it can take on large initiatives, which can save a significant amount of money. There is always an opportunity to reduce waste and optimize rates.

Motivating engineering teams to take action

Engineering teams are highly focused on delivering features with speed and agility so the business can capture new markets, attract new customers, improve efficiency, and find new revenue streams. FinOps teams require attention and commitment to bring the visibility of spending, but these initiatives are often treated as low priority by engineering teams. In the *Operate* phase, FinOps teams establish processes so the visibility they gain can be converted either into cost savings or optimized cloud spending.

The FinOps team has two approaches to handle the engineering team's inability to commit. The first approach is to incentivize engineering teams who commit to initiatives and act in time to produce the desired results. The second approach is to scrutinize or penalize teams for not acting on FinOps recommendations and producing the desired results. It is always recommended to start with a positive approach first and then move on to a negative one.

Incentivizing the team

Let's take an example. *Team A* and *Team B*, both spend $500,000/month on cloud costs. FinOps has identified a potential opportunity to reduce costs by right-sizing virtual machines and utilizing the weekend shut-down schedule. Through the business case, the FinOps team has clearly shown the amount each team could save by implementing these initiatives. *Team A* understands this priority and commits to taking care of cost optimization work items in their sprint planning and starts implementing it.

The FinOps team has created a dashboard showing the progress of both teams. As *Team A* moves through the cost optimization recommendations and implements them, their month-over-month spend is reduced by 30% and it is clearly displayed on the dashboard. The dashboard is visible to the CEO and the progress of each team is visible.

Since *Team A* has realized savings by taking action, it is rewarded by the organization showcasing their achievements during a town hall event, increasing the morale of all who committed the time and effort.

Penalizing the team

In the above example, *Team B* was given the opportunity to commit to and act on the cost savings initiative but they did not act on it or dedicate time and effort toward it. The dashboard, which is visible across the organization, lists *Team B* under the "worst offender" list. The list serves as a reminder that *Team B* still has pending things to do. All engineering teams take pride in the work they do. Being on the worst offender list, the organization can scrutinize the team's budget increase request, ask them to justify the current spend, and control the finances until positive actions are taken toward the initiative that FinOps has started.

Next, let's examine how implementing automation for various tagging activities can enhance compliance efforts.

Automated tag inheritance, governance, and compliance

Azure tags are widely used to allocate the cost to business units and engineering departments. Azure Cost Management has a feature, which when enabled, automatically applies the subscription or resource group tags to child resources so that you don't have to individually tag them. The Tag inheritance feature is available to Enterprise Agreement customers and those with Microsoft customer agreement accounts.

Here is an example of how tag inheritance works. The resource gets both the resource group and subscription tags applied to it.

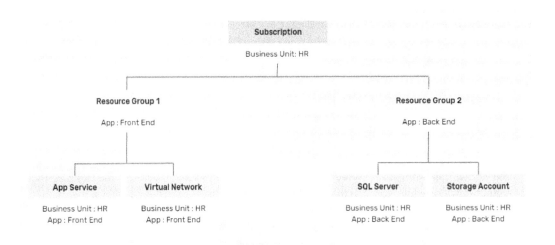

Figure 9.1 – Tag Inheritance structure

To enable automated tag inheritance, do the following:

1. Open the Microsoft Edge browser.

2. Navigate to `https://portal.azure.com` and sign in to your organization's account.

3. In the top search bar, search for `Cost Management + Billing` and select the highlighted service.

4. Under **Settings**, click on **Manage Subscription**.

5. Click on **Edit** under **Tag inheritance (preview)**. If you do not see this feature, make sure to click the **Try Preview** button and enable all features.

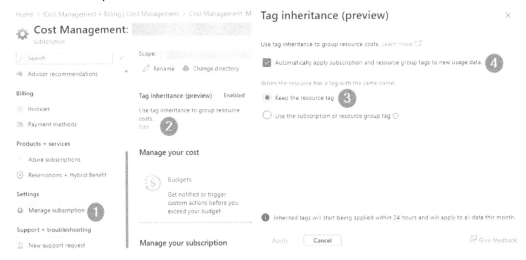

Figure 9.2 – Tag Inheritance settings

6. Check the box to automatically apply tag inheritance, select the **Keep the resource tag** radio button, and click **Apply**.

You can use Azure cost analysis to view the costs grouped by tags as well; you can download your Azure usage and charges report and it will contain the inherited tag details.

Even though you have automated tag inheritance in place, you still may want to enforce certain tags to be present when a subscription, resource group, or resource is created. For example, an organization can have a mandatory tag policy to identify the business unit and owner. In this case, you can use Azure Policy to enforce that the tag exists and deny the resource creation if it's not present.

To create an Azure policy for required tags, do the following:

1. Open the **Microsoft Edge** browser.
2. Navigate to `https://portal.azure.com` and sign in to your organization's account.
3. In the top search bar, search for `Policy` and select the highlighted service.
4. Click on **Assignments** and then **Assign policy**.
5. Under the **Basics** tab, click on **Policy definition**.
6. Search for `Require a Tag` and select that policy.

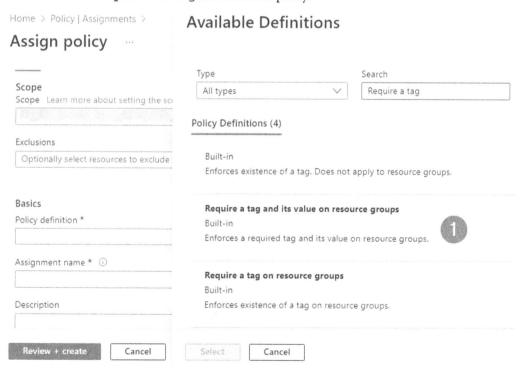

Figure 9.3 – Required tag policy

7. Fill in the rest of the details and click **Review + create**.

8. Now, once this policy is in place and users try to create a new resource without the required tag, they will get the following error message.

Figure 9.4 – Policy violation message

Azure policies are an excellent way to enforce and govern not only tags – there are a wide variety of capabilities that can be leveraged by FinOps teams.

Continuing our exploration of automation, let's now consider scheduling virtual machine startup and shutdown.

Automated VM shutdown and startup

Automated VM shutdown and startup is an excellent way to save money on dev/test workloads for virtual machines as well as workloads that are predictable. There are approximately 730 hours in a month. If you shut down one virtual machine for Saturday and Sunday, that will reduce the billable hours by 26%. That 26% is a straight saving achieved by automated VM shutdown. Auto shutdown based on schedules has been around for quite some time, but the lack of auto startup features has limited its adoption – until now...! Azure now offers the ability to stop and start a VM at a scheduled time of the day.

To configure auto stop and start for a virtual machine, do the following:

1. Open the Microsoft Edge browser.

2. Navigate to `https://portal.azure.com` and sign in to your organization's account.

3. In the top search bar, search for `Virtual Machines` and select the highlighted service.

4.   From the list of virtual machines, select the one you want to set up an auto shutdown and start schedule for.

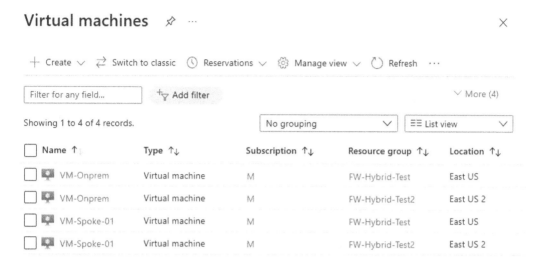

Figure 9.5 – List of virtual machines

5.   Click on **Tasks** and then click on **Add task**.

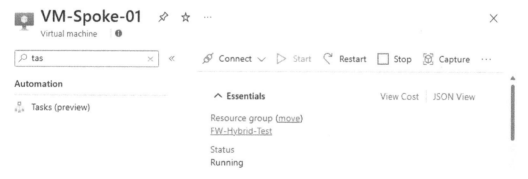

Figure 9.6 – Automation tasks

6.   Click on the **Power off Virtual Machine** task.

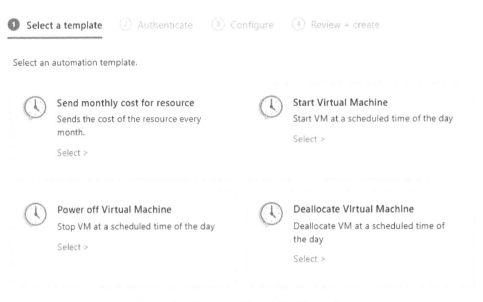

Figure 9.7 – Power off Virtual Machine task

7. Next, create a link under **Connections** to authenticate to Azure VM and Office 365. Then click on the **Next: configure** button.

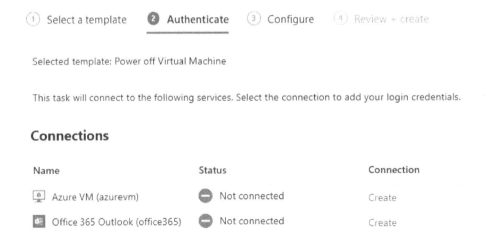

Figure 9.8 – Connection authentication

8. Fill out the task details based on your requirements. In this example, we want to shut down the development machine at 9 p.m. daily.

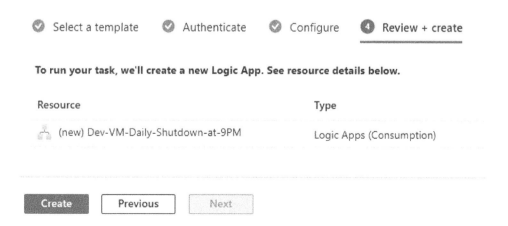

Figure 9.9 – Shutdown task schedule

9. Optionally, you can enable VM shutdown email notifications.

10. Next, click on **Review + create**.

Figure 9.10 – Automatic VM shut down task details

As you will have noticed, the tasks use consumption-based Azure logic apps to perform the VM shutdown as scheduled.

You can now go back to **Tasks** and you will see the shutdown task has been created. Click on **Add a Task** to configure a new auto startup for a virtual machine:

1. Select the **Start Virtual Machine** task.

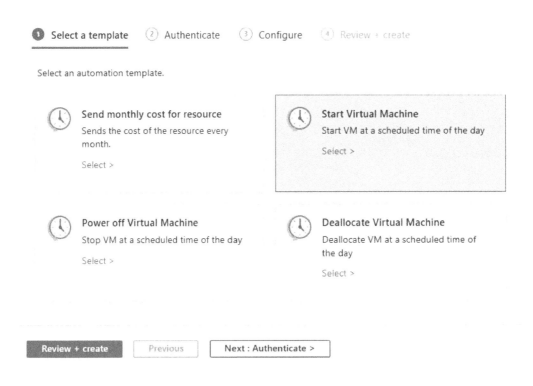

Figure 9.11 – Start Virtual Machine task

2. Go to **Next : Authenticate** and you should see your account is already linked (as part of the previous steps).

3. Click on **Configure** and create an auto-start schedule as follows:

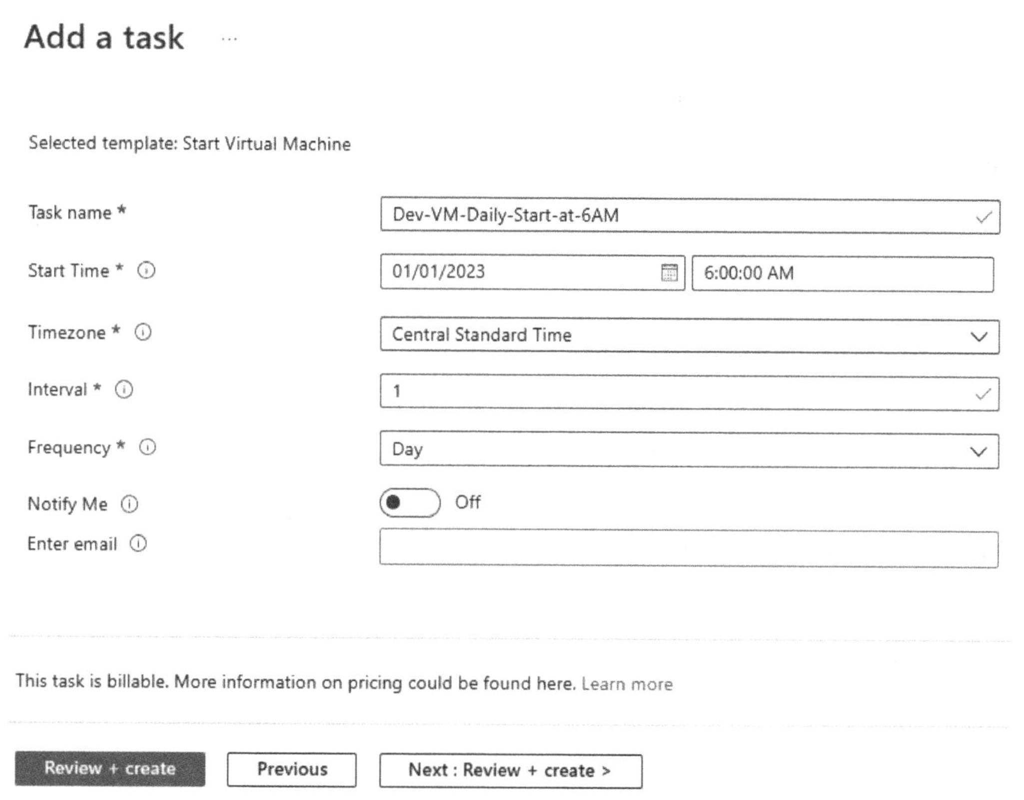

Add a task   ⋯

Selected template: Start Virtual Machine

Task name *	Dev-VM-Daily-Start-at-6AM ✓
Start Time * ⓘ	01/01/2023 📅	6:00:00 AM
Timezone * ⓘ	Central Standard Time ⌄
Interval * ⓘ	1 ✓
Frequency * ⓘ	Day ⌄
Notify Me ⓘ	⬤ Off
Enter email ⓘ	

This task is billable. More information on pricing could be found here. Learn more

[Review + create]   [Previous]   [Next : Review + create >]

Figure 9.12 – Start VM schedule

4. Click **Review + create** to create the logic app to start the virtual machine at 6 a.m. daily.

> **Note**
> Shutdown and startup tasks are based on consumption-based logic apps. Thus, these tasks are billable.

Let's now turn our attention to budget action automation, which can help manage costs by automatically taking action when specific cost thresholds are exceeded.

Automated budget actions

Azure budgets provide a mechanism to notify users as well as to trigger an action to prevent future costs by automating shutting down a VM, scaling in a VM scaleset, and reducing Request Units in Cosmos DB.

Automated budget actions are suitable for any of the following scenarios where turning off or reducing the number of VMs does not critically impact business continuity:

- Resources in dev/test subscriptions
- CI/CD virtual machine scale sets
- Resources in sandbox subscriptions
- Resources in R&D environments

There are various ways a budget trigger can kick off remediation using an Azure Monitor action group. We will use the automation runbook method, but you can use any one of these options:

- Automation runbook
- Azure function
- Logic app
- Secure webhook
- Webhook

To create an automated budget action using a runbook, do the following:

1. Open the Microsoft Edge browser.
2. Navigate to `https://portal.azure.com` and sign in to your organization's account.
3. In the top search bar, search for `Automation Account` and click **Create**.
4. Fill in the basic details and keep the rest of the settings as the defaults.

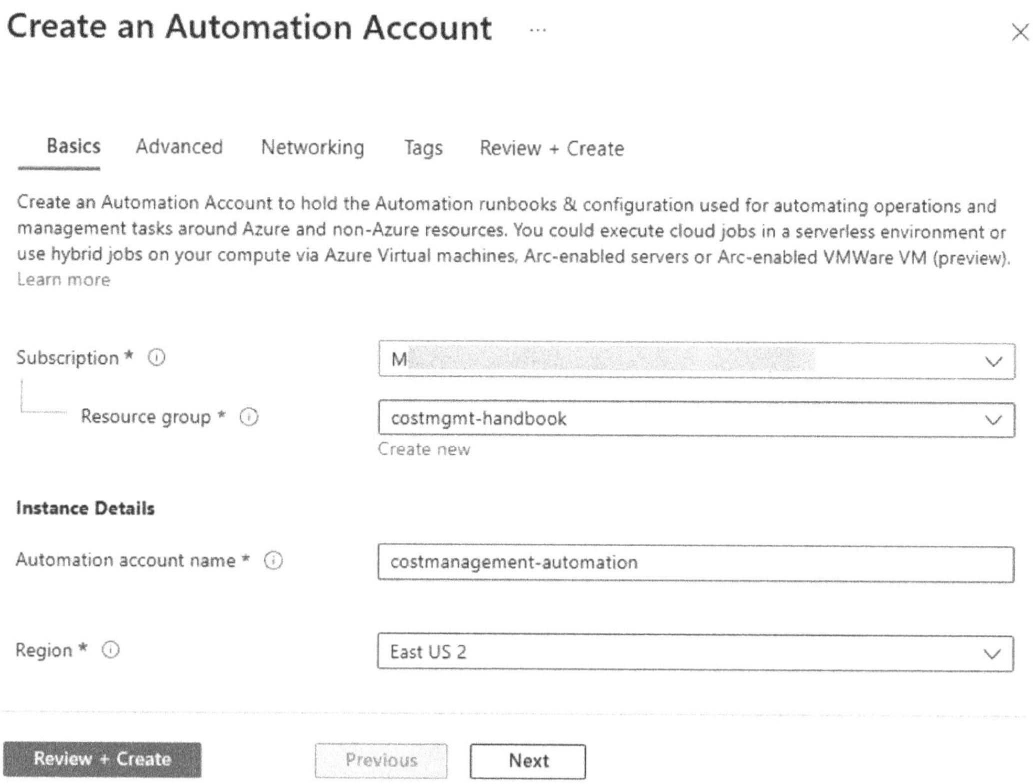

Create an Automation Account ...

×

Basics   Advanced   Networking   Tags   Review + Create

Create an Automation Account to hold the Automation runbooks & configuration used for automating operations and management tasks around Azure and non-Azure resources. You could execute cloud jobs in a serverless environment or use hybrid jobs on your compute via Azure Virtual machines, Arc-enabled servers or Arc-enabled VMWare VM (preview). Learn more

Subscription * ⓘ

M

Resource group * ⓘ

costmgmt-handbook

Create new

Instance Details

Automation account name * ⓘ

costmanagement-automation

Region * ⓘ

East US 2

Review + Create   Previous   Next

Figure 9.13 – Create an automation account

5. Click **Review + Create**.

6. Once the automation account is created, go to the account, click on **Runbooks**, and then click on **Create new runbook**.

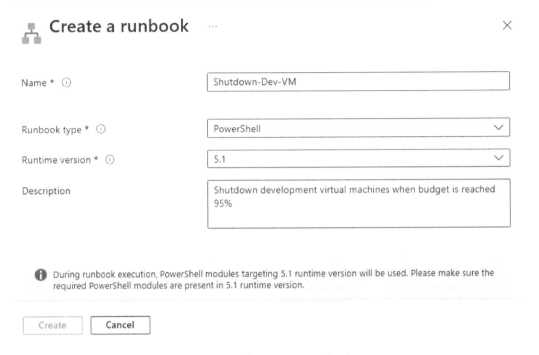

Figure 9.14 – Create a new runbook

7.   Click **Create**. Copy and paste the following PowerShell code into the editor:

```
$myconnectionName = "AzureRunAsConnection"
try
{
    $servicePrincipalConnection=Get-AutomationConnection
-Name $myconnectionName
    "Logging to Azure"
    Connect-AzAccount `
        -ServicePrincipal `
        -TenantId $servicePrincipalConnection.TenantId `
        -ApplicationId $servicePrincipalConnection.
ApplicationId `
        -CertificateThumbprint
$servicePrincipalConnection.CertificateThumbprint
}

#Shutdown only VMs where budgetshutdown tag is yes
$VMs = Get-AzVm  |  Where {$_.Tags.Keys -contains
```

```
"budgetshutdown" -and $_.Tags.Values -contains "yes"} |
Select Name, ResourceGroupName, Tags
ForEach ($VM in $VMs)
{
    $VMStatus = Get-AzVM -Name $VM.Name
-ResourceGroupName $VM.ResourceGroupName -Status |
    Select Name, ResourceGroupName, DisplayStatus, Tags
    $VMN=$VM.Name
    $VMRG=$VM.ResourceGroupName
        If ($VMStatus = "VM Running")
            {
                Stop-AzVM -Name $VMN -ResourceGroupName
$VMRG -force
                "$VMN is now Shutdown and Deallocated due
to budget threshold reached." }}
```

8. Click **Save** and then click **Publish**.

Figure 9.15 – PowerShell script to stop VM based on tag

Next, we will create an action group in Azure Monitor and connect our VM shutdown PowerShell workbook to the action group:

1. In the top search bar, search for `Monitor` and click on **Alerts** in **Azure Monitor**.

Figure 9.16 – Azure Monitor Alerts

2. Click on **Action groups** to create a new action group.

3. Fill In the **Basics** tab information and click on **Actions**. Select **Automation Runbook** for **Action type**. In the **Configure Runbook** window, make sure to set **Runbook source** as **User** to find the runbook we just created.

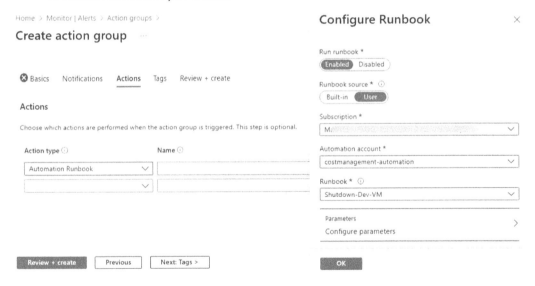

Figure 9.17 – Create an action group

4. Next, click **Review + create**.

At this point, we have our automation runbook and action group created. Next, we need to configure the action group in the budget alert configuration:

1. Go to **Azure Cost Management** and click on **Budget** to create a new budget.

2. In the **Set alerts** tab, under **Alert conditions**, select **Actual**, **95** for **% of budget**, and **shutdown-dev-vm** for **Action group**.

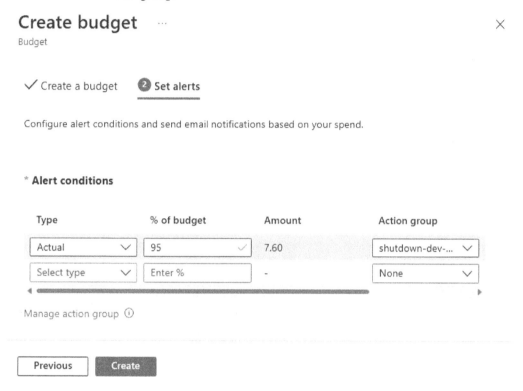

Figure 9.18 – Set budget threshold

3. Click **Create**.

4. Now when the budget exceeds the threshold, it will trigger the alert and call the Azure automation workbook to shut down all the VMs tagged with the budgetshutdown = yes tag.

 Microsoft Azure

You have an alert for budget 'shutdown-vm'

Your total spend for budget 'shutdown-vm' is now $12.01, exceeding your specified threshold value of $7.20.

Budget name	shutdown-vm
Budget start date	January 1, 2023
Budget type	Cost
Budget value	$8.00
Actual value	$12.01
Notification threshold	$7.20

View in Azure portal >

Figure 9.19 – Email alert notifying VM shutdown action

Budget alerts are evaluated every 12 hours. So, if you don't see the action running, wait until the next day to confirm it shut down the virtual machines.

Third-party FinOps tools

The FinOps team often must decide whether they should build the necessary reports, dashboards, and tooling or buy a product that provides those capabilities. It's a classic build versus buy decision. If your team is new and does not have time to develop tooling, buying third-party products is often a good idea. These products have been on the market for quite some time and are mature enough to have all the necessary aspects of FinOps built into them.

Apptio Cloudability

Cloudability by Apptio allows FinOps teams to establish a budget and allows them to accurately forecast and track spending in Azure. It provides a current month, forecast, and budget view. Business leaders can correlate cloud spending to business value using cost allocation. It also provides anomaly and right-size recommendation reports to FinOps teams to further enhance savings. Cloudability has a nice feature called **Reserved Instance Planner**, which provides recommendations for purchasing new Reserved Instances and exchanging existing Reserved Instances for better coverage.

For more information, please visit `https://www.apptio.com`

CloudHealth by VMware

VMware offers a cloud cost management product called CloudHealth. It provides advanced resource and organization management to view and analyze cloud costs by project, line of business, application, and cost center. It comes with canned cost reports as well as allowing you to customize it based on your needs. In terms of budgeting, it provides a budget view of actual, budget, and variance month-over-month reports. Apart from cost allocation and chargeback capabilities, it also provides commitment-based discount management. CloudHealth has advanced reservation purchase, optimization, and amortization capabilities to manage discounts throughout purchases.

For more information, please visit `https://cloudhelath.vmware.com`

Cast.ai

If you are looking for a Kubernetes-focused all-in-one platform then Cast.ai offers cost management, optimization, and automation. You can connect to your AKS clusters and see recommendations for autoscaling, spot instance savings, bin packing and right-sizing, dynamic node types, and cluster hibernation. With this tool, you get AKS namespace, workload, and cluster-level cost reporting. In addition to cost management, it also provides cluster issue reports, vulnerability reports, and security best practice recommendations.

For more information, please visit `https://cast.ai`

Let's summarize what we learned in this chapter.

Summary

In this chapter, we looked at how to establish a Center of Excellence for cloud cost management. Building a FinOps team and hiring and training the right resources are key to its success. Cloud costs are a very vast and dynamic field and new features are constantly being added. A centralized FinOps team is well suited to tackle the challenges of cost optimization and savings. Once processes are established by the CoE, the FinOps team uses automation to handle tag inheritance, automated VM shutdown schedules and automated budget actions to remediate over-budget resources.

Let's address the elephant in the room in the next chapter: how to allocate cost for the most popular, yet challenging, technology for FinOps – containers.

10

Allocating Costs for Containers

In this chapter, we will explore how to allocate costs for container workloads. Microservices have gained popularity, and running containers in **Azure Kubernetes Service** (**AKS**) poses challenges in allocating costs for showback and chargeback purposes. Shared AKS clusters bring another challenge. We will offer a brief overview of **Azure Container Instances** (**ACI**) and AKS from a tagging perspective. Since Azure's native cost allocation capability for AKS is very limited, we will explore the open source tool called Kubecost. FinOps teams rely on Kubecost, and it has become an industry standard for in-cluster, external, and shared cost allocation and visibility. We will also examine the showback and chargeback mechanisms using Kubecost.

In this chapter, we will cover the following topics:

- FinOps challenges for containerized workloads
- ACI cost allocation
- Introducing Kubecost
- AKS cost allocation
- Showback and chargeback for shared AKS clusters
- Cost optimization recommendations for AKS clusters

Let's get started!

Technical requirements

We will be using the following tools to accomplish the tasks in this chapter:

- The **Kubecost** open source tool, available at `https://github.com/kubecost`

- An existing AKS cluster

- Microsoft's **Cost Management + Billing** tool, available at `https://portal.azure.com/#view/Microsoft_Azure_CostManagement/Menu/~/overview`. Alternatively, you can also find the Cost Management + Billing tool by signing into the Azure portal and, in the top-center search bar, typing `Cost Management + Billing`.

When using these tools, sign in to the Azure portal using your organization ID. Please refer to *Chapter 1* for the minimum Role Based Access Control permissions required to use Azure **Cost Management + Billing**.

FinOps challenges for containerized workloads

In today's world, it is becoming extremely common to containerize a workload and host it in the cloud using AKS. AKS is a managed service that provides many advanced features and ease of deployment. An organization with a sizable AKS usage may have many clusters for each development, testing, and production environment. Along with the workload clusters, there may exist shared clusters. Shared AKS clusters host workloads from multiple teams and for multiple end customers.

The *tagging* strategy that works perfectly for all other cloud resources does not effectively work for AKS. The main reason why tagging does not work for AKS is the sheer complexity of abstracting the underlying resources. For example, when you create an AKS cluster, it creates an `MC_` management resource group. All the dependent resources, such as virtual machines, storage accounts, and virtual networks, are in a managed resource group. Now, since the single AKS cluster will host multiple containers for multiple teams, it does not make sense to tag the Azure resources to a single cost center or development team. We have to tag the individual containers that are running inside the cluster. Currently, Azure does not provide tagging at the container level; thus, we have to rely on open source tools such as Kubecost to get the cost allocation details from within the cluster.

While container orchestrators such as AKS cause challenges, in single container instance services such as the ACI service, it is possible to tag an individual service. Since ACI can run only a single container workload, it can be tagged to a single cost center. Azure container registries can be tagged to the same cost center, or their cost can be shared by all the application teams as a shared resource. Let's look at how to allocate cost for ACI.

ACI cost allocation

ACI and container apps both support tags. For single container workloads hosted on these platforms, we can create *department*, *business unit*, *cost center*, and *owner* tags. These tags then will be available in the Billing data export file as well as in the Cost Management tool. Also, remember to tag Azure Container Registry, as it is a related service and is used to deploy containers in Azure Container Instances.

To allocate costs for ACI, follow these steps:

1. Open the **Microsoft Edge** browser.
2. Navigate to `https://portal.azure.com` and sign in with your organization's account.
3. In the top search bar, search for `Container Services` and select the highlighted service.
4. Select the existing container service you have.
5. Click on **Tags** on the **Overview** page of your ACI instance.
6. Add the **Department**, **Business Unit**, **Cost Center**, and **Owner** tags.

Edit tags   ✕

Tags are name/value pairs that enable you to categorize resources and view consolidated billing by applying the same tag to multiple resources and resource groups. Tag names are case insensitive, but tag values are case sensitive. Learn more about tags ↗

Tags

Name ⓘ		Value ⓘ	
Department	:	HR	🗑
Business Unit	:	BU01	🗑
Cost Center	:	C12290	🗑
Owner	:	Jane Doe	🗑
	:		

Resource

🔵 vote-front (Container instances)
    4 to be added ⓘ

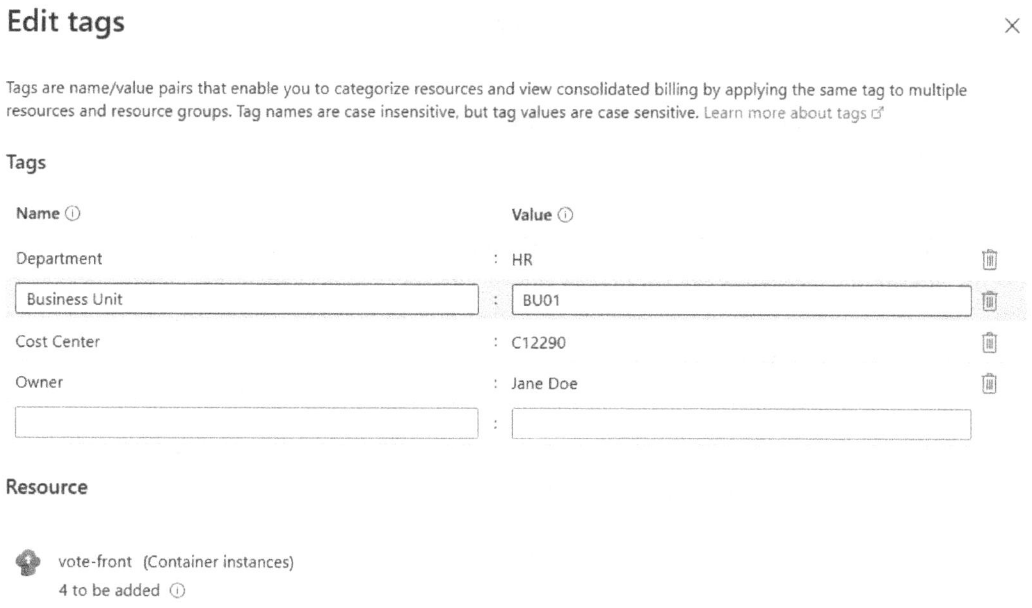

Figure 10.1 – Tagging an Azure container instance

7. Click **Save** to apply the tags.

8. Next, open an Azure Container Registry instance and apply the **Department**, **Business Unit**, **Cost Center**, and **Owner** tags. Since Azure Container Registry is a resource related to our Azure Container Instances, be sure to use the same cost center to show the cost properly in the billing reports.

Figure 10.2 – Tagging Azure Container Registry

9. Open Azure Cost Management and click on **Cost analysis**.

10. Next, click on **Group By** and select **Tag | Cost Center | C12290**.

11. Cost Management now shows both the costs of ACI and Azure Container Registry, based on the selected cost center tag.

Next, we will examine cost allocation for AKS, which does not support namespaces and container-level tags natively. To address this limitation, we will explore Kubecost, a third-party solution for AKS cost allocation.

Introducing Kubecost

Kubecost (`https://github.com/kubecost`) started in early 2019 as an open source project to give developers visibility of how much they spend on Azure Kubernetes Service. Since then, it has become a de facto standard to allocate and monitor the cost of AKS. Kubernetes environments are challenging, and shared AKS clusters are a nightmare for FinOps teams to allocate the `cost` property to, due to many in-cluster and out-of-cluster service dependencies.

Kubecost features cost allocation, unified cost monitoring, optimization insights, and alerts and governance. We will be focusing on the cost allocation and cost monitoring aspects of Kubecost in this chapter. If you want to learn more about cost optimization or the alert and governance side of Kubecost, please visit `https://docs.kubecost.com` for more information.

We will also configure the Azure cost report export to view out-of-cluster costs (e.g., Azure SQL Database or a storage account) and to reconcile costs with our actual cloud bill to reflect enterprise discounts, spot market prices, and commitment discounts. In addition to this, the report also allows us to configure negotiated discounts to reflect the billing amount as closely as possible.

The AKS cluster has been pre-configured with the Azure Voting App, which can be accessed at `https://github.com/Azure-Samples/azure-voting-app-redis`. However, any other multi-container application hosted in your AKS would also be suitable for the purposes of this exercise.

AKS cost allocation

Cost allocation for AKS is complex, and Azure currently does not support detailed cost allocation (by namespace, node, container, deployment, etc.). Due to this, we will be using Kubecost to get the cost allocation capability.

First, we will configure Kubecost to our existing AKS cluster, and then we will configure the Azure billing export so that Kubecost can read the billing data and learn about out-of-cluster costs, as well as our rates and discounts.

To configure Kubecost, follow these steps:

1.   Open the **Microsoft Edge** browser.

2.   Navigate to `https://portal.azure.com` and sign in with your organization's account.

3.   In the top search bar, search for `Kubernetes Services` and select the highlighted service.

4.   From the list of AKS clusters, click on the cluster you want to install Kubecost.

5.   Next, click on **Connect** to get the instructions to connect to your AKS cluster.

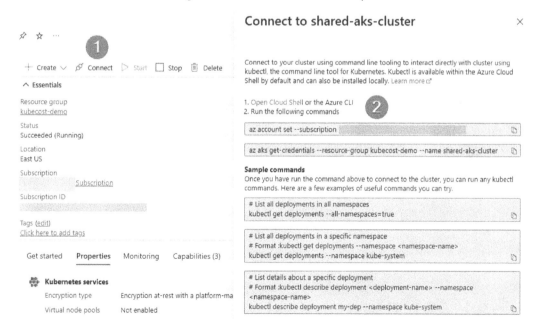

Figure 10.3 – Connecting to the AKS cluster

6. Open Command Prompt on your computer and follow the steps described in the **Connect to shared-aks-cluster** window. If you run into issues, make sure you have the Azure CLI installed and it's in the PATH (`https://learn.microsoft.com/en-us/cli/azure/install-azure-cli`)

7. Open a new tab in the browser, visit `https://www.kubecost.com`, and click on **Get Started**.

Kubecost is an open core tool that provides visibility into Kubernetes spend and resource allocation. It leverages this data to provide capacity management insights, avoid outages, and reduce costs.

To gain access, enter your email below:

myemail@myorg.com

Get Started

Figure 10.4 – Getting started with Kubecost

8. Provide your email address.

9. Next, you will see the page with instructions on how to install Kubecost.

10. Back in Command Prompt, run the following commands:

```
kubectl create namespace kubecost
helm repo add kubecost https://kubecost.github.io/cost-analyzer/
helm install kubecost kubecost/cost-analyzer --namespace kubecost --set kubecostToken="bXllbWFpbEBteW9yZy5jb20=xm343yadf98"
```

11. Next, to view the Kubecost UI, run the following command to enable the port forwarding:

```
kubectl port-forward --namespace kubecost deployment/
kubecost-cost-analyzer 9090
```

12. Open a new tab in the browser and type `http://127.0.0.1:9090`:

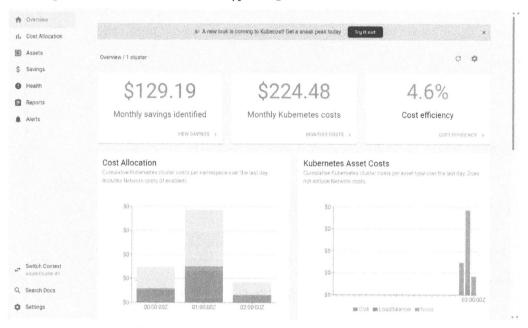

Figure 10.5 – The Kubecost overview page

13. You will see the Kubecost UI.

Now that we have Kubecost installed, let's configure the Azure out-of-cluster cost using the billing report exported into the storage account. The complete documentation is located here: `https://docs.kubecost.com/install-and-configure/install/cloud-integration/azure-out-of-cluster`:

1. Go to **Azure Cost Management** and click on **Exports** under **Settings**.

2. Click on **Schedule export**.

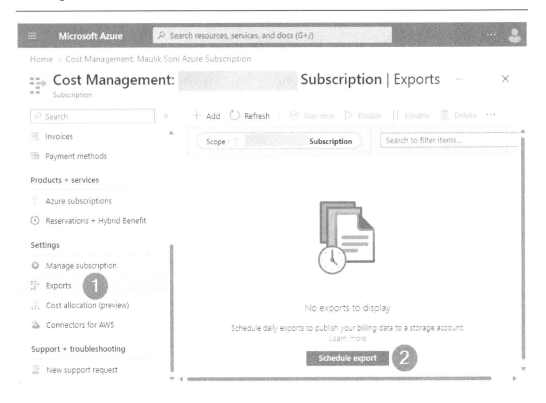

Figure 10.6 – Scheduling the export of a billing report

3.   For **Metric**, select **Actual cost (Usage and Purchases)** from the dropdown, and for **Export type**, select **Daily export of month-to-date costs**. Leave file partitioning off, select your storage account, and then click **Create**.

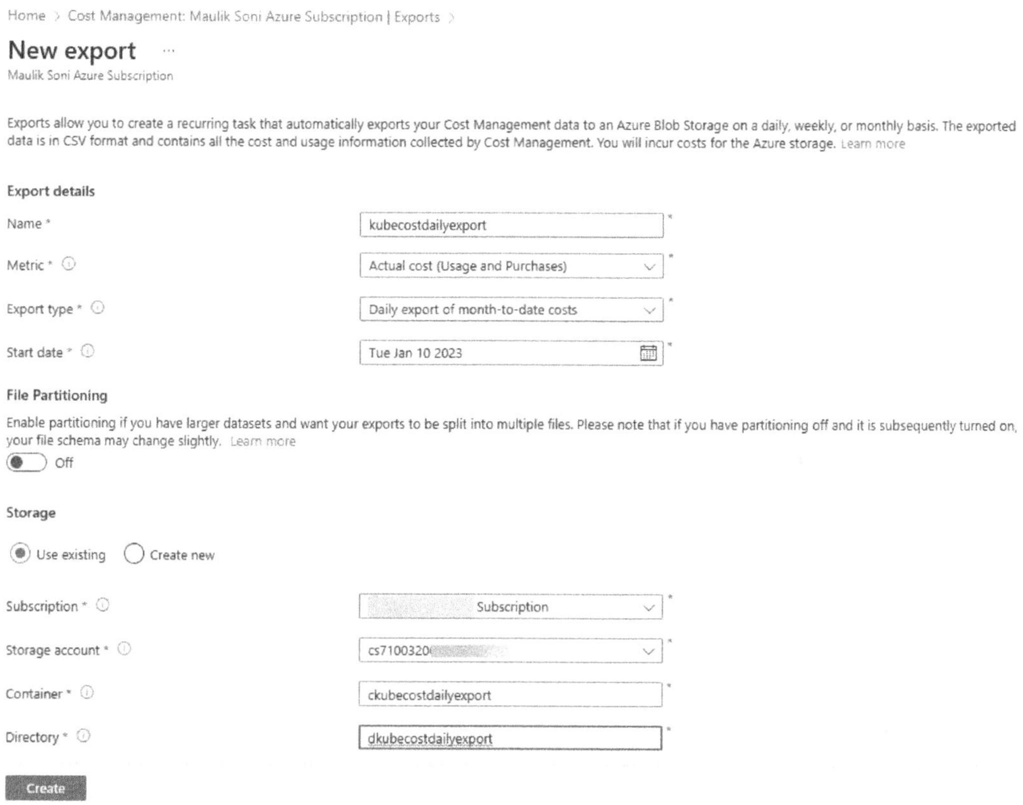

Home > Cost Management: Maulik Soni Azure Subscription | Exports >

New export   ···
Maulik Soni Azure Subscription

Exports allow you to create a recurring task that automatically exports your Cost Management data to an Azure Blob Storage on a daily, weekly, or monthly basis. The exported data is in CSV format and contains all the cost and usage information collected by Cost Management. You will incur costs for the Azure storage. Learn more

Export details

Name *	kubecostdailyexport
Metric * ⓘ	Actual cost (Usage and Purchases) ∨
Export type * ⓘ	Daily export of month-to-date costs ∨
Start date * ⓘ	Tue Jan 10 2023 🗓

File Partitioning

Enable partitioning if you have larger datasets and want your exports to be split into multiple files. Please note that if you have partitioning off and it is subsequently turned on, your file schema may change slightly. Learn more

(●) Off

Storage

(●) Use existing   ( ) Create new

Subscription * ⓘ	Subscription ∨
Storage account * ⓘ	cs7100320 ∨
Container * ⓘ	ckubecostdailyexport
Directory * ⓘ	dkubecostdailyexport

Create

Figure 10.7 – Configuring the billing data export details

4. After 24 hours, verify that the Azure billing data export is available in the storage account.

5. Next, we need to provide Kubecost access to the billing CSV files.

6. Go back to the Kubecost portal at `http://127.0.0.1:9090/settings`, locate the **Cloud cost settings** section, and click **UPDATE** on the external cloud cost configuration item.

7.   In the pop-up window, provide the storage account details.

Figure 10.8 – The Azure billing data export configuration in the Kubecost UI

8.   Next, follow the steps to configure Azure Rate Card API integration at `https://docs.kubecost.com/install-and-configure/install/cloud-integration/azure-out-of-cluster/azure-config`.

9.   Once both tasks are completed, navigate to the **Kubecost Assets** view, and it will no longer show a banner that says **External cloud cost not configured**.

Now that we have configured the out-of-cluster cost and billing rate, let's explore Kubecost from a cost allocation perspective.

Showback and chargeback shared AKS clusters

In economics, the tragedy of the commons is characterized by individuals having unrestricted access to resources, resulting in overuse and depletion to the detriment of society. This same issue can be seen with shared computing resources, with application owners believing that cloud infrastructure is limitless and practically free. This can lead to over-provisioning and underutilization, which is why there needs to be a shared understanding of accountability among the finance department, DevOps, software developers, and application owners.

Implementing a chargeback program for an AKS cluster requires the following steps, shown in *Figure 10.9*:

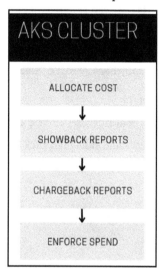

Figure 10.9 – The chargeback program steps

While we have talked about allocating costs for an AKS cluster using Kubecost, we haven't yet discussed showback and chargeback. In the following example, we have an AKS cluster hosting two workloads – **Azure-Vote** and **Kubecost**.

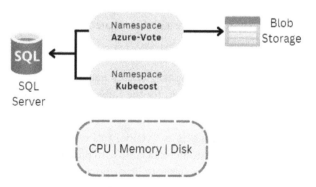

Figure 10.10 – A shared AKS cluster and its resources

Showback reporting does not require financial systems to transfer funds and generate financial statements as a chargeback program does. Despite this, it still gives cost center managers the ability to view their infrastructure usage trends over time.

In the preceding example, the database is a shared resource. It's being utilized by both namespaces. The blob storage is considered an external cost only for the Azure-Vote workload. The AKS cluster resources (CPU, memory, disk, networking, etc.) are considered in-cluster costs.

Now, using Kubecost, we can generate a showback report by following these steps:

1. Open the **Microsoft Edge** browser.

2. Navigate to the Kubecost home page at `https://localhost:9090`.

3. Click on the **Reports** link on the left-hand side, then click the **Create a Report** dropdown, and select **Allocation Report**.

Figure 10.11 – The Kubecost UI showing the cost allocation report

4. For the time range, select **Month-to-date** and **Aggregated by Namespace**.

5. Click **Save Report** and save it as `Showback Report for shared cluster A01`.

In the preceding showback report, we can see the cost for each namespace. The cost of running Azure-Vote is around $1.51 a month to date, and the cost for the Kubecost workload is around $1.97. Now, let's look at what chargeback is.

In a chargeback model, the IT department acts as a vendor to a cost center, which is a customer of theirs. Like in business, the cost center/customer can choose to purchase their infrastructure needs from the IT department, or a third-party supplier with lower prices and better products and services. Senior management will assess the profit and loss of each business unit in this model, requiring the cost center to set budgets, forecast spending, and make decisions based on the numbers. The chargeback model is simply a step on the path to creating and enforcing a culture of cost-effective infrastructure spending across a company.

We will need to add a few labels to the namespace to relate it to the cost center. To add the required labels, follow these steps:

1. Open Command Prompt.

2. Connect to your existing Azure AKS cluster by following the instructions provided in the Azure portal.

3. Once connected, run `label` commands, as follows:

    ```
    kubectl label namespaces azure-vote department=D001
    kubectl label namespaces azure-vote team=FE001
    kubectl label namespaces azure-vote product=P001
    kubectl label namespaces azure-vote env=prod
    ```

4. Once the labels are applied, go to **Allocations** and **Aggregate by Department** to view the department chargeback amount.

Figure 10.12 – The Kubecost UI showing the chargeback amount using the Department label

5. Click on **Save Report** and name it `Chargeback Report for department D001`.

Saved reports enable filtering, grouping, and sorting, allowing us to report various types of metadata. These saved reports also support cost allocation decisions, such as choosing to share idle costs by node or cluster and deciding how to account for a cluster's shared resources, either evenly distributing them or weighing the distribution by cost.

Apart from cost allocation, Kubecost also provides cost optimization recommendations. Let's look at those next.

Cost optimization recommendations for AKS clusters

In *Chapter 5*, *Hitting the Goals for Usage Optimization*, in *Workbook 10 – Azure Kubernetes Services*, we looked at how to reduce AKS costs by using a cluster auto scaler, spot VMs, and start/stop AKS cluster strategies. Since Kubecost provides granular visibility into an AKS cluster, let's look at some recommendations it offers. To view these recommendations for your AKS cluster, go to the Kubecost home page (`https://localhost:9090`), and then click on the **Savings** tab.

Manage underutilized nodes

Nodes in an AKS cluster refer to the Virtual machine instances that provide compute for the cluster. To view whether the nodes are overprovisioned, click on the **Manage underutilized nodes** report in the **Savings** tab.

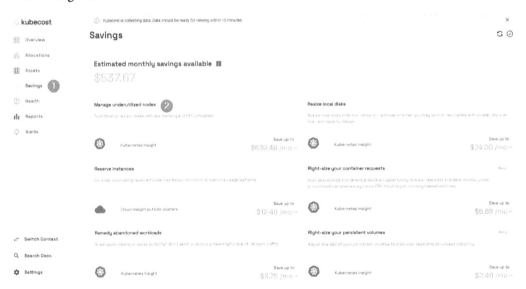

Figure 10.13 – Kubecost showing savings

From the **Savings** list, let's select **Manage underutilized nodes** to see more details.

Cluster Savings

Nodes with underutilized CPU & memory

Nodes with low memory and CPU utilization are candidates for being turned down or resized. The following nodes have sustained usage below 25% in both categories. Your cluster has enough resource availability to support turning these nodes down.

Maximum CPU/RAM Request Utilization (60%)

Node	Node Checks	Pod Checks	Recommendation	
aks-agentpool-20532064-vmss00000k	Passed	Passed	Safe to drain. Save $106.58 / mo.	↓
aks-agentpool-20532064-vmss00000n	Passed	Passed	Safe to drain. Save $106.58 / mo.	↓

Figure 10.14 – The underutilized nodes recommendation

The **Cluster Savings** page will provide a list of nodes with a **Node Checks** status (**Passed** or **Failed**) and a recommendation to safely drain the node with expected savings. In this example, we have provisioned eight nodes, but the sustained memory and CPU utilization is for only two nodes. We can drain the six nodes from the node pool and realize $639 of monthly savings.

Resizing local disks

Attached disks with under 20% current usage and 30% predicted usage can often be resized to save costs.

Cluster Savings   ↻ ⊘

‹ Back to savings

Local Disks with low utilization

Attached disks with under 20% current usage and 30% predicted usage can often be resized to save costs. Consider launching new nodes with smaller disks on the next node turndown.

Disk Name	Cluster	Current Utilization	Recommendation	Savings	Actions
aks-agentpool-20532064-vmss00000g ⌝	cluster-one	15.7%	Resize disk to 75 Gb	$3.00	⋯
aks-agentpool-20532064-vmss00000j ⌝	cluster-one	15.7%	Resize disk to 75 Gb	$3.00	⋯
aks-agentpool-20532064-vmss00000m ⌝	cluster-one	17.2%	Resize disk to 75 Gb	$3.00	⋯
aks-agentpool-20532064-vmss00000h ⌝	cluster-one	15.8%	Resize disk to 75 Gb	$3.00	⋯

Figure 10.15 – The Local Disks with low utilization report

In the preceding example, current disc utilization is between 15% and 17%, and it recommends resizing the disc to 75 Gb to realize total savings of $25 a month.

Reserved instances

The **Reserved Instances** recommendations look at historical resource usage patterns and calculate the savings.

Reserved Instances

⟲ ⊘

‹ Back to savings

A recommended strategy for purchasing reserved instances is to observe the sustained usage/request low point (1st percentile) for both RAM and CPU capacity. This is often a good indicator for your future floor for CPU/RAM needs.

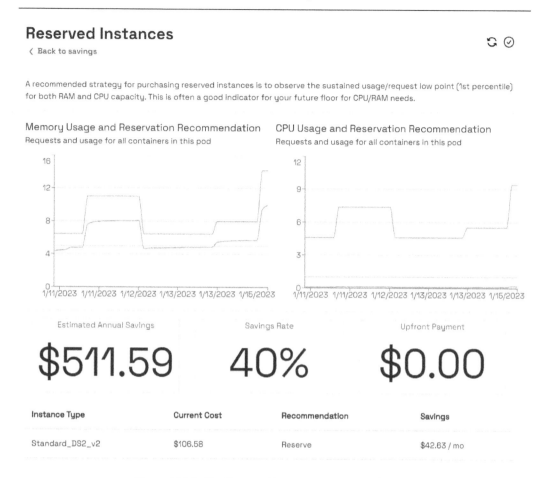

Figure 10.16 – The Reserved Instances recommendations

In the preceding example, we have eight standard DS2_v2 VMs, and by purchasing reservations, we can realize 40% savings annually compared to a pay-as-you-go cost.

Right-sizing your container requests

Over-provisioned containers provide an opportunity for fewer requests and saving money. Under-provisioned containers may cause CPU throttling or memory-based evictions. It is important to right-size containers by looking at historical data.

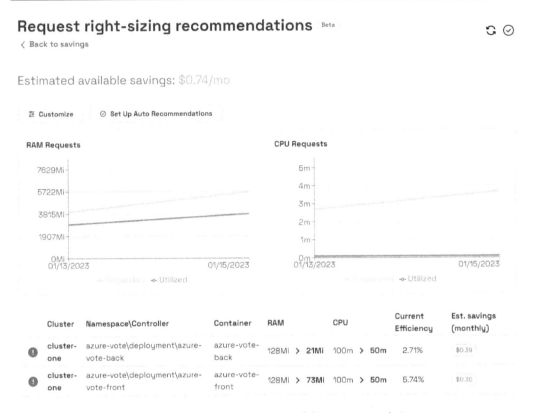

Request right-sizing recommendations `Beta`

‹ Back to savings

Estimated available savings: $0.74/mo

⇄ Customize   ⊘ Set Up Auto Recommendations

RAM Requests

CPU Requests

	Cluster	Namespace\Controller	Container	RAM	CPU	Current Efficiency	Est. savings (monthly)
❗	cluster-one	azure-vote\deployment\azure-vote-back	azure-vote-back	128Mi › 21Mi	100m › 50m	2.71%	$0.39
❗	cluster-one	azure-vote\deployment\azure-vote-front	azure-vote-front	128Mi › 73Mi	100m › 50m	5.74%	$0.35

Figure 10.17 – The right-sizing RAM and CPU recommendations

In the preceding example, we first have to filter the namespace to `azure-vote` to select our workload and view the right-size recommendations. The current CPU and RAM allocation efficiency is low, between 2% and 5%. Thus, we have overprovisioned. Resize the RAM from 128 Mi to 21 Mi, and the CPU from 100 m to 50 m.

Remedying abandoned workloads

Scale down, delete, or resize pods that don't send or receive a meaningful rate of network traffic. You can set the traffic threshold and duration and measure it against the workload.

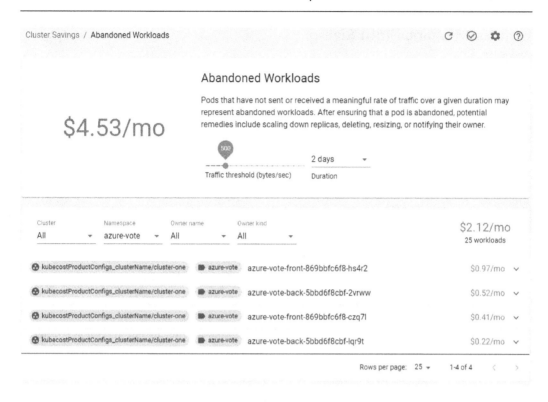

Figure 10.18 – Abandoned Workloads recommendations

In the preceding example, we have selected 500 bytes per second as a threshold and usage duration for 2 days. In addition to that, we have filtered the namespace to the **azure-vote** workload. The detailed view shows that our application has ingress and egress activities at 0 bytes per second. Based on this observation, we can advise the engineering team to consider scaling down a replica, deleting, or resizing the pod.

Right-size persistent volumes

Kubernetes pods are temporary and expendable resources, which means that applications often require persistent storage. In Azure, this storage is supported by Azure Disks, Azure Files, Azure NetApp Files, or Azure Blob. Identifying and optimizing the size of non-utilized persistent volumes is crucial for ensuring efficient cost optimization.

Figure 10.19 – The Persistent Volume Right Sizing recommendation

In the preceding example, we have two recommendations to reduce the current persistent storage allocation from 34.4 GB to 1 GB, based on the maximum usage pattern.

Now, let's summarize what we have learned in this chapter!

Summary

In this chapter, we looked at the cost allocation challenges for containerized workloads, due to the complexity of identifying the in-cluster cost, the external cost, and the shared cost. Container orchestrators such as AKS abstract the inner workings, which makes it impossible to tag all resources with cost centers. The answer to the challenge is Kubecost, an open source, real-time cost visibility and insights tool that provides the needed cost allocation visibility, by running a container inside an AKS cluster and monitoring various container and runtime metrics. We also learned how to configure Kubecost for an existing AKS cluster. Using the cost allocation features of Kubecost, we learned how to use data for showback and chargeback reporting. And finally, we looked at the savings recommendations that Kubecost provides and the actions engineering teams can take to realize the savings.

In the next chapter, we will explore metric-driven cost optimization.

11
Metric-Driven Cost Optimization

Metric-Driven Cost Optimization (**MDCO**) is a cost-management strategy for cloud computing environments. It involves using data and analytics to continuously monitor, measure, and optimize cloud costs. By using metrics, organizations can gain a deep understanding of their cloud spend and identify areas where they can reduce costs.

This approach involves collecting and analyzing data on cloud resource utilization, cost, and performance. The data is used to understand the organization's cloud spending patterns and to identify areas for optimization. Based on this information, organizations can make informed decisions about their cloud infrastructure investments, such as reducing the use of underutilized resources or selecting more cost-effective resources.

MDCO can also help organizations to better understand their cloud spend over time, track their spending against budgets, and identify areas where they can reduce costs. It enables organizations to be proactive in their cost management, rather than simply reacting to cost spikes. By continuously monitoring and optimizing cloud costs, organizations can improve their financial performance, reduce the risk of overspending on cloud resources, and make the most of their investment in cloud computing.

In this chapter, we will learn about the following:

- Core principles of MDCO
- MDCO and reservation reporting using Power BI
- Setting thresholds for purchasing reservations
- Automated reservation purchases based on MDCO triggers

Let's get started!

Technical requirements

We will be using the following tools to accomplish the tasks in this chapter:

- Microsoft's Cost Management + Billing tool, available at `https://portal.azure.com/#view/Microsoft_Azure_CostManagement/Menu/~/overview`. Alternatively, you can also find Cost Management + Billing by signing in to the Azure portal and, in the search bar, typing `Cost Management + Billing`.

- Microsoft Power BI Cost Management Connector, available at `https://learn.microsoft.com/en-us/power-bi/connect-data/desktop-connect-azure-cost-management`.

- A Microsoft Power BI Pro or Premium license is required to use the data-driven alert feature.

- A SQL Server database with read and write permissions.

- A Power Automate per-user license and deployment environment: `https://powerautomate.microsoft.com/`.

When using these tools, sign in to the Azure portal using your organization ID. Please refer to *Chapter 1* for the minimum RBAC permissions required to use Azure **Cost Management + Billing**.

Core principles of MDCO

Let's look at the core principles of MDCO:

- **Automated measurement**: The continuous and automated measurement of cloud resource utilization, cost, and performance to identify and measure optimizations

- **Data-driven decision-making**: MDCO relies on data and analytics to inform decision-making and identify areas for cost reduction

- **Resource optimization**: Optimizing the use of cloud resources, such as reducing the use of underutilized resources or selecting more cost-effective resources, is a key component of MDCO

- **Proactive cost management**: MDCO enables organizations to be proactive in their cost management rather than simply reacting to cost spikes

- **Budget tracking**: MDCO enables organizations to track their spending against budgets and identify areas where they can reduce costs

- **Cost transparency**: MDCO provides organizations with greater transparency into their cloud spend, enabling them to make more informed decisions about their cloud infrastructure investments

- **Continuous improvement**: MDCO is an ongoing process, and organizations should continuously monitor and optimize their cloud costs to ensure that they are making the most of their investment in cloud computing

MDCO can be applied to usage optimization to reduce wastage, as well as rate optimization to purchase reservations. In this chapter, we will focus on how to purchase reservations based on the MDCO process.

MDCO and reservation reporting using Power BI

MDCO depends on near real-time monitoring of data and making decisions supported by the key characteristics of the data. When purchasing a reservation, it is often overwhelming to simulate various scenarios to make the optimal decision. One such scenario is to compare optimal savings between 1-year and 3-year reservations for Azure services. FinOps teams are often asked to create such reports to allow businesses to evaluate the various choices.

Here is an example MDCO recreation Power BI report that you can customize to suit your needs:

1. Open Microsoft Edge.

2. Navigate to `https://github.com/PacktPublishing/FinOps-Handbook-for-Microsoft-Azure/tree/main/Chapter-11`.

3. Download `ri-savings-pbi.pbix` and open it in Power BI Desktop.

Metric Driven Cost Optimizations for Reservations

Region	RI_Category	Service_Instance	Term	Breakeven_point_Months	Cur^
eastus2	ManagedDisk	Premium_SSD_Managed_Disks_P30	P1Y	12	
eastus2	MariaDB	AzureDB_MariaDB_General_Purpose_Compute_Gen5	P1Y	7	
eastus2	MySQL	AzureDB_MySQL_General_Purpose_Compute_Gen5	P1Y	7	
eastus2	sqldatabases	SQLDB_BC_Compute_Gen5	P1Y	8	
eastus2	sqldatabases	SQLDB_GP_Compute_Gen5	P1Y	8	
eastus2	sqldatabases	SQLDB_HyperScale_Compute_Gen5	P1Y	8	
eastus2	sqldatabases	SQLMI_GP_Compute_Gen5	P1Y	8	
eastus2	virtualmachines	Standard_B16ms	P1Y	7	
eastus2	virtualmachines	Standard_B1ms	P1Y	8	

Figure 11.1 – MDCO for reservations report

4. The graphic in the top left is **Term** selection. The **RI Savings** data has a field that can be very useful for distinguishing reservation savings between 1-year and 3-year commitments. In the example, we have selected 1-year commitment.

5. The second graphic is the **Annual Cost vs. Estimated Savings** section when we purchase the reservations. This is a very important metric because it clearly communicates the amount and percentage of savings. In the example, if we purchase recommended reservations for a 1-year term, we will be saving around $74K per year, which is 38% of the annual cost.

6. The third visual is the **Breakeven** point. It is helpful to understand when our upfront payment for reservation will be equal to the pay-as-you-go rate. In our example, for our 1-year term, we have a range of breakeven points falling between 7 months and 12 months. It is recommended to purchase the reservations with the lowest breakeven point to realize the savings quickly.

7. The following table in the report shows most of the key data. This table is interactive. When you switch the term from 1 year to 3 years, all the graphic and table data is updated.

There are lots of opportunities to create different reports to suit your organization's needs. The example report we have walked through will help you to start the journey.

Setting thresholds for purchasing reservations

Organizations in the crawl stage of FinOps tend to purchase reservations on a schedule basis, that is, quarterly or semi-annually. For example, a FinOps analyst who works with the engineering team identifies the virtual machine SKUs that are being used consistently and proposes to purchase reservations. The engineering team then purchases the reservations, and the process ends until the next quarter comes, when it repeats.

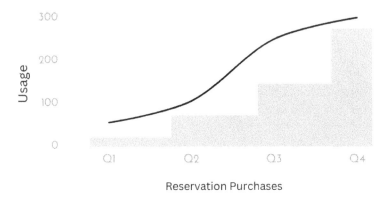

Figure 11.2 – Quarterly reservation coverage

Schedule-based reservation purchasing has some downsides. When the workload is constantly evolving, having a fixed schedule to buy reservations results in coverage gaps. This coverage gap is essentially a loss of savings opportunities. The team could have purchased reservations more frequently if they had access to the right usage data at the right time.

Alternatively to schedule-based purchases, the FinOps team can set up a dashboard that shows the eligible virtual machine usage that is currently not covered by the reservation. The dashboard then can define a threshold. The threshold could be any valid numerical value; when crossed, it triggers a notification.

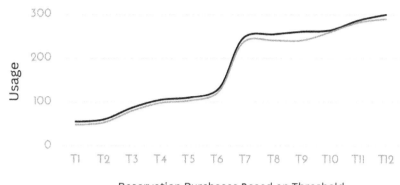

Reservation Purchases Based on Threshold

Figure 11.3 – Reservation coverage with a threshold to trigger a purchase

Once the FinOps team receives the threshold trigger for a reservation purchase, they work with the engineering team to complete the purchase. Note that the time gap between trigger T1 and T2 could be days, weeks, or months. Essentially, FinOps and engineering teams will not act unless they receive the threshold trigger to purchase new reservations.

Here is a scenario. The FinOps team has been asked to build a Power BI report to show Azure reservation recommendations and, if the potential savings are greater than $170K, the report should send an automated email alert to notify the engineering team.

This is how you can accomplish this:

1. Open Microsoft Edge.

2. Navigate to `https://portal.azure.com` and sign in to your organization's account.

3. In the search bar, search for `SQL Server` and select the SQL Server instance that we created in *Chapter 3, Forecasting the Future Spend*.

4. Select your SQL database and click on **Query Editor**.

5. For this exercise, we will be using `RI Recommendation` sample data. Please download the `https://github.com/PacktPublishing/FinOps-Handbook-for-Microsoft-Azure/blob/main/Chapter-11/ri-savings-insert-script1.txt` file from GitHub. Alternatively, you can also use the Azure `cost management` Power BI Connector described in *Chapter 3, Forecasting the Future Spend*, and query the *RI_Recommendations* table.

6. Copy the SQL statements from the `ri-savings-insert-script1.txt` file, paste them into the SQL database query editor, and click **Run**.

Figure 11.4 – SQL Server query editor view

7. Next, on the left menu, click on **Power BI** and click on **Get started**.

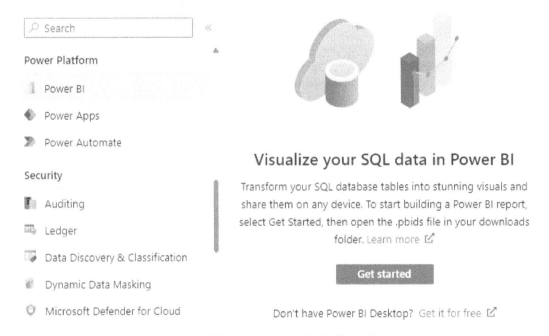

Figure 11.5 – Getting started with Power BI

8. The browser will download the `<<databasename>>.pbids` file.

9. Open the Power BI Desktop, navigate to your `Downloads` folder, and open the `pbids` file that we just downloaded.

10. Select **ri_savings** in the **Navigator** window and click **Load**.

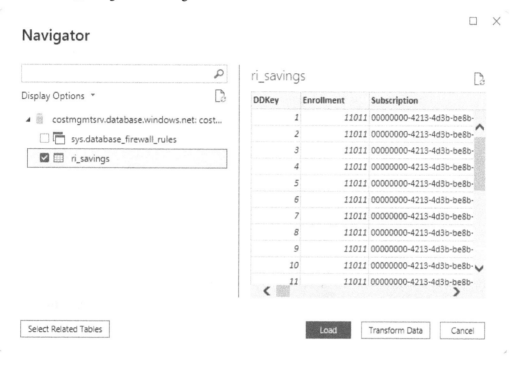

Figure 11.6 – Navigator window in Power BI Desktop

11. In the **Power BI Reports** view, click on **Table visualization** and select the **Subscription, Resource_Id**, **Region**, **Estimated_RI_Savings**, **RI_Category**, and **Service_Instance** fields.

12. Next, add a gauge visualization and select **Estimated_RI_Savings**.

ption	Resource_Id	Region	Sum of Estimated_RI_Savings	RI_Category	Service_Instance
a80aed					
)00-4213- ie8b- a80aed	aee329a80aed	eastus2	4,815.44	virtualmachines	Standard_D16as_v
)00-4213- ie8b- a80aed	aee329a80aed	eastus2	923.54	virtualmachines	Standard_D2_v2
)00-4213- ie8b-	aee329a80aed	eastus2	11,143.72	virtualmachines	Standard_D2s_v3
			168,802.12		

Figure 11.7 – Gauge visualization in our Power BI report

13. In order to configure data alerts, we need to publish this report to Power BI Online.

14. Click on the **Publish** button in Power BI Desktop, select your workspace, and click **Next**.

15. Once the report is published, click on the **Open** in Power BI Link.

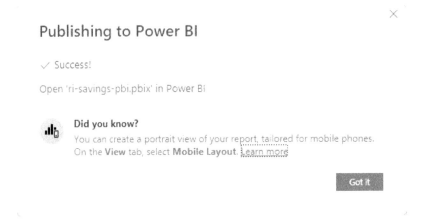

Figure 11.8 – Publishing our Power BI report

16. Next, pin the **Estimated_RI_Savings** Gauge visualization to the dashboard to enable the data alerts, then go to the dashboard.

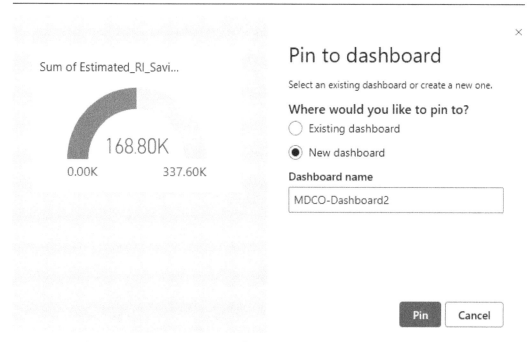

Figure 11.9 – Pinning report visualization to the dashboard in Power BI Online

17. In the dashboard, click on three dots and select **Manage alerts**.

Figure 11.10 – Managing alerts for our pinned visual

18. Add an alert rule and configure the alert condition. Set **Threshold** to 170000 and then click on **Save and close**.

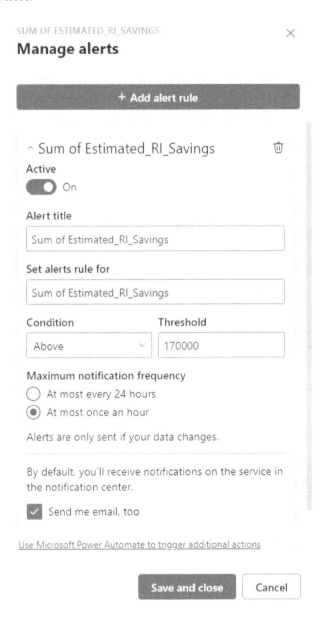

Figure 11.11 – The Manage Alerts window

Now, let's add some more reservation recommendations to our SQL table, which should trigger a notification email to purchase the reservations.

19. Please download the `https://github.com/PacktPublishing/FinOps-Handbook-for-Microsoft-Azure/blob/main/Chapter-11/ri-savings-insert-script2.txt` file from GitHub.

20. Go back to the SQL editor in the Azure portal, copy the SQL `insert` statements from the `ri-savings-insert-script2.txt` file, paste them in the editor, and click on **Run**.

21. After an hour, when the dataset refreshes, it will update the **Estimated_RI_Savings** gauge to cross the $170,000 threshold and trigger an alert.

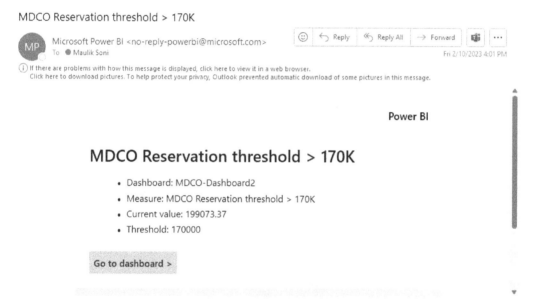

Figure 11.12 – Example email alert

If you would like to learn more about Power BI's data alerting feature, please take a look at the Microsoft documentation here: `https://learn.microsoft.com/en-us/power-bi/create-reports/service-set-data-alerts`.

In this section, we configured automated alerts based on a threshold value. When the FinOps analyst receives the alerts, they validate it with the engineering team and schedule a reservation purchase. But can we automate the reservation purchase as well? Let's find out!

Automated reservation purchases based on MDCO triggers

Previously, we learned how to send an alert email alert based on an MDCO trigger to look at potential reservation purchases. It is certainly possible to automate a reservation purchase based on the same trigger. Here is a diagram showing what the automated flow could look like:

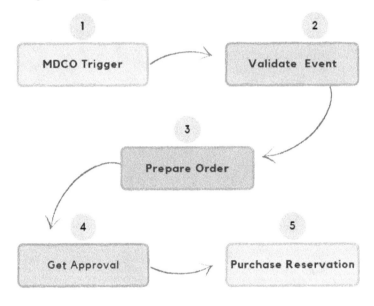

Figure 11.13 – MDCO automation steps

Here are the steps to achieve the automation described in the preceding diagram:

1. Subscribe to the MDCO trigger event to get notified when the pre-defined threshold is reached.

2. Validate the MDCO trigger by querying the backend database. This step ensures there is a legitimate event that will kick off the reservation purchase flow.

3. Prepare the order by querying the backend database and identifying the Azure service, SKU, region, term, payment, and scope of the reservation order.

4. Request an approval email flow to send out notifications and approval requests. This step ensures the finance, FinOps, and engineering teams are aware of the type of reservation being automatically purchased.

5. Finally, submit the reservation purchase order using the Azure Reservations REST API.

When the reservations are purchased successfully, Azure will automatically notify the reservation owners and reservation administrators.

> **Note**
>
> A Microsoft Power Automate per-user license is required to complete this exercise. Due to organization-specific tooling and complexities, the implementation of Steps 2, 3, and 4 discussed in the preceding list is outside the scope of this book. Azure Functions may be used to perform the business logic check before the purchase. Please check out the Microsoft documentation to learn more about Power Automate: `https://learn.microsoft.com/en-US/training/paths/automate-process-power-automate/`.

Now, let's implement the solution using Microsoft Power Automate:

1. Open Microsoft Edge.

2. Navigate to `https://make.powerautomate.com/` and sign in with your account.

3. Click on **Create** and select **Automated Cloud Flow**.

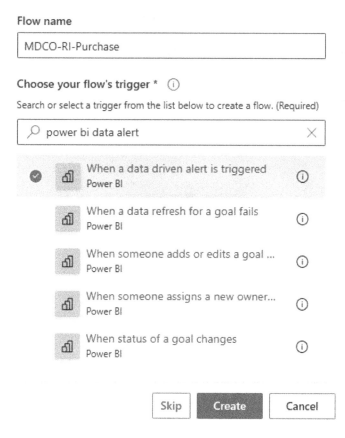

Figure 11.14 – Create a new automation flow

4. Type in a flow name, then search for `Power BI data alert` and select **When a data driven alert is triggered**.

5. In the flow designer, select **Alert Id** as follows:

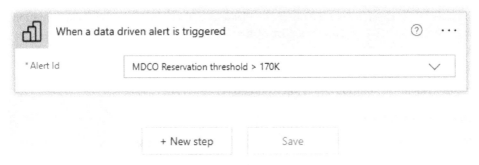

Figure 11.15 – First step of the automation flow

6. Click on + **New step**, search for `HTTP with Azure AD`, and select **Invoke an HTTP request**.

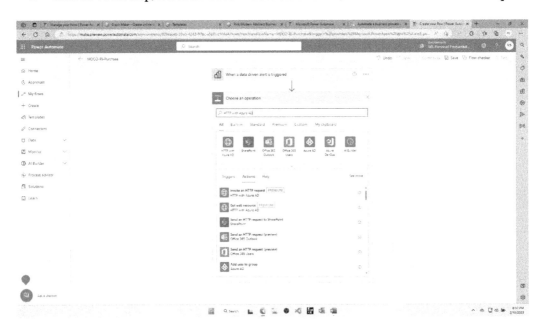

Figure 11.16 – Second step of the automation flow

7. Next, type https://management.azure.com in the **Base Resource URL** field and the **Azure AD Resource URI (Application ID URI)** field:

Figure 11.17 – HTTP with Azure AD connection details

8. Click **Sign in**.

9. Next, we will be making an HTTP REST API request to purchase the reservations. The **Body of the request** field has all the details about the reservation SKU, region, payment, and so on:

Figure 11.18 – REST API request to purchase reservations

10. Replace the {subscription-id}, {reservation-id}, and {product-id} placeholders with appropriate values and click **Save** at the top right to save the flow.

Automation allows the timely purchase of Azure reservations by mounting and analyzing RI Recommendations data.

Now, we are at the end of the chapter. Let's summarize what we have learned so far.

Summary

In conclusion, cloud FinOps is an essential aspect of managing costs in the cloud, and MDCO is a powerful tool for achieving this goal. By leveraging metrics and data analysis, organizations can gain insights into their cloud spending, identify inefficiencies, and make data-driven decisions to optimize their cloud costs. This approach can help organizations control their cloud costs while maximizing the value of their cloud investments, leading to improved efficiency, productivity, and profitability. With the right tools and strategies, any organization can leverage cloud FinOps and MDCO to achieve their cost management goals and drive success in the cloud.

Unit economics is a fundamental concept in business that examines the costs and revenues associated with a single unit of a product or service. Let's explore it in the next chapter.

12

Developing Metrics for Unit Economics

Unit economics in cloud FinOps refers to the analysis of the costs and revenue associated with delivering a single unit or product within the cloud services offered by a business. The purpose of unit economics analysis is to understand the financial performance of these individual units, and how changes in costs can impact the overall profitability of the business. This information is used by FinOps teams to make informed decisions about pricing, cost optimization, and resource allocation, as well as identify opportunities for improving efficiency, reducing waste, and minimizing costs. In short, unit economics is a critical aspect of cloud FinOps, providing valuable insights for improving the financial performance of cloud-based businesses.

In this chapter, we will learn about the following:

- What is cloud unit economics?

- Indirect versus direct cost metrics

- Tracking costs back to business benefits

- Developing metrics for unit economics

- Activity-based cost model

Let's get started!

Technical requirements

We will be using the following tool to accomplish the tasks in this chapter:

- Microsoft **Cost Management + Billing**, available at `https://portal.azure.com/#view/Microsoft_Azure_CostManagement/Menu/~/overview`. Alternatively, you can also find **Cost Management + Billing** by signing in to the Azure portal and typing `Cost Management + Billing` in the top-center search bar.

When using this tool, sign in to the Azure portal using your organization ID. Please refer to *Chapter 1* for the minimum RBAC permissions required to use Azure **Cost Management + Billing**.

During this chapter, we will use the healthcare industry as an example to illustrate how to create metrics for unit economics. Let's start with an introduction to cloud unit economics.

What is cloud unit economics?

Cloud unit economics is a system that measures the cost and usage metrics associated with dynamic infrastructure changes in cloud computing. It applies the concept of unit economics to the cloud computing domain and calculates the difference between marginal cost and marginal revenue to determine where cloud operations break even and begin to generate profit. While cloud unit economics is often discussed in the context of commercial SaaS and cloud-driven commercial organizations, it can also be applied to non-profit or public sector entities by replacing the terms *profit* or *revenue* with *value*.

Benefits of cloud unit economics

The main benefits of using cloud unit economics are as follows:

- Forecasting profitability and understanding factors affecting margins
- Building a plan for cloud cost optimization
- Evaluating a product's future potential and making changes to the product roadmap and engineering priorities
- Responsible use of cloud resources by end users
- Contribution of cloud engineers to gross profit margin and supporting pricing decisions
- Better forecasting of cloud costs despite variable consumption

Let's look at the distinction between direct and indirect costs for services to develop **key performance indicators** (**KPIs**).

Indirect versus direct cost metrics

When developing unit metrics for a service, it is important to consider both **indirect** cost metrics (which refer to the costs incurred prior to production) and **direct** cost metrics (which are incurred during production). Indirect cost metrics include costs associated with **research and development (R&D)**, marketing, and other non-production expenses. Direct cost metrics, on the other hand, include costs associated with cloud resources, labor, and other expenses that are directly tied to the production of the service.

Choosing the appropriate cost metrics to measure can depend on the specific goals of the organization. For example, a company that is focused on improving its overall profitability may want to focus on reducing indirect costs to improve its margins. Alternatively, a company that is seeking to improve the quality of its services may want to focus on reducing direct costs and improving the efficiency of its production processes.

For example, healthcare organizations can develop the following indirect cost metrics:

- Total cost of sandbox subscriptions (R&D)
- Total cost of non-production subscriptions (R&D)
- Forecasted future cost of new scheduling system

These are the direct cost metrics:

- Total cost of production subscriptions
- Cost per patient encounter
- Revenue per patient encounter
- Cost per claim
- Seasonal forecasted cost to serve patients

Ultimately, a combination of both indirect and direct cost metrics is necessary to obtain a comprehensive view of the financial health of a service. By carefully measuring and monitoring these metrics, organizations can identify areas for improvement and make data-driven decisions to optimize their operations and maximize profitability.

Let's understand how we can track costs back to the business benefits.

Tracking costs back to business benefits

Tracking costs back to business benefits from a unit economics perspective involves understanding the cost and revenue associated with each unit of product or service provided by a healthcare organization and using this information to optimize business decisions.

To achieve this, healthcare organizations need to develop a clear understanding of their unit economics by identifying the cost of providing each unit of service, and the revenue generated from each unit. This information can be used to calculate KPIs such as cost per patient encounter, revenue per patient encounter, and gross margin per patient encounter.

By tracking these KPIs over time, healthcare organizations can identify which services or patient segments are the most profitable, and where there may be opportunities to increase profitability by reducing costs or increasing revenue. This information can then be used to optimize business decisions, such as pricing strategies, resource allocation, and investment in new services or technologies.

In the context of cloud FinOps, tracking costs back to business benefits from a unit economics perspective involves understanding how cloud spending impacts the cost and revenue associated with each unit of service provided by a healthcare organization. This can be achieved by mapping cloud spending to specific business units or projects and using cloud cost allocation and chargeback mechanisms to associate cloud costs with the specific units of service that they support. This information can then be used to optimize cloud spending, improve unit economics, and, ultimately, drive better business outcomes for the healthcare organization.

Let's look at how to develop the metrics for unit economics next.

Developing metrics for unit economics

To enhance our comprehension of the unit economic metrics, let us conceptualize the unit metrics for a healthcare organization. However, it should be noted that collecting, cleaning, and modeling data is beyond the scope of this book due to the sheer complexity and data regulation of healthcare organizations.

Cost per patient encounter

Cost per patient encounter (CPE) is a KPI used to measure the cost of providing healthcare services for each patient encounter in a healthcare organization.

To calculate the CPE, healthcare organizations need to identify all the costs associated with providing care to a patient during a single encounter. This may include costs related to cloud resources, such as storage, compute, and network usage, as well as other costs such as the facility, labor, equipment, and supplies.

Once these costs have been identified, the total cost can be divided by the number of patient encounters to calculate the CPE. For example, if a healthcare organization spends $10,000 on cloud resources and other costs to provide care for 1,000 patient encounters, the CPE would be $10 per patient encounter.

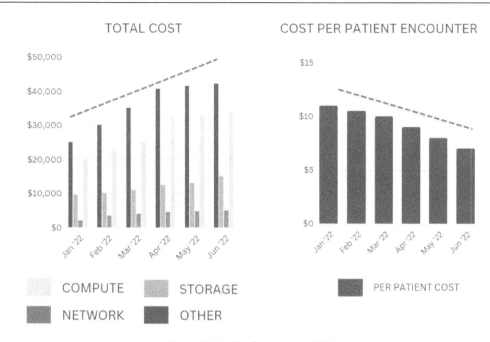

Figure 12.1 – Total cost versus CPE

In the preceding example, the total cost is going up over time but the CPE is going down due to an increased number of patient visits. By tracking the CPE over time, healthcare organizations can identify trends and make informed decisions about where to allocate resources to improve the financial performance of their healthcare services.

Revenue per patient encounter

Revenue per patient encounter (**RPE**) is a KPI used to measure the revenue generated from each patient encounter.

To calculate the RPE, healthcare organizations need to identify all the revenue generated from each patient encounter. This may include revenue from insurance reimbursements, copayments, and other fees.

Once the total revenue has been identified, it can be divided by the number of patient encounters to calculate the RPE. For example, if a healthcare organization generates $20,000 in revenue from insurance reimbursements, copayments, and other fees for 1,000 patient encounters, the RPE would be $20 per patient encounter.

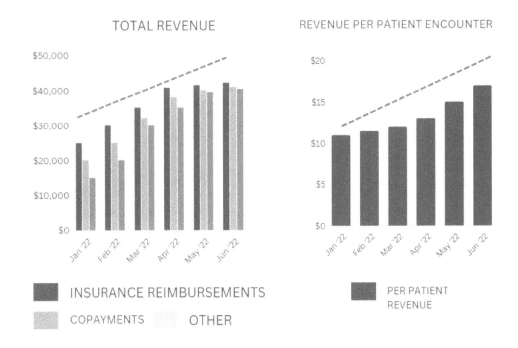

Figure 12.2 – Total revenue versus RPE

By tracking the RPE over time, healthcare organizations can also identify trends and make informed decisions about which services to prioritize and how to optimize their pricing strategy to improve the financial performance of their healthcare services.

Gross margin per patient encounter

Gross margin per patient encounter (GMPE) is a KPI used to measure the profitability of providing healthcare services for each patient encounter.

To calculate the GMPE, healthcare organizations need to identify the revenue generated from each patient encounter and subtract the total costs associated with providing care during the encounter, including costs related to cloud resources such as storage, compute, and network usage, as well as other costs such as facilities, labor, equipment, and supplies.

Once the revenue and cost have been identified, the difference can be divided by the number of patient encounters to calculate the GMPE. For example, if a healthcare organization generates $20,000 in revenue from insurance reimbursements, copayments, and other fees for 1,000 patient encounters, and the total cost to provide care during those encounters was $10,000, the GMPE would be $10 per patient encounter.

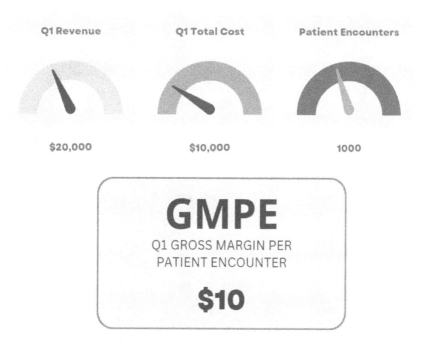

Q1 Revenue   Q1 Total Cost   Patient Encounters

$20,000 $10,000   1000

GMPE

Q1 GROSS MARGIN PER
PATIENT ENCOUNTER

$10

Figure 12.3 – Q1 GMPE

By tracking the GMPE over time, healthcare organizations can also identify trends and make informed decisions about which services to prioritize and how to optimize their pricing and cost structure to improve the financial performance of their healthcare services.

Cost per claim

Cost per claim (**CPC**) is a KPI used to measure the cost of processing healthcare insurance claims.

To calculate the CPC, healthcare organizations need to identify all the costs associated with processing healthcare insurance claims, including costs related to cloud resources such as storage, compute, and network usage, as well as other costs such as facilities, labor, equipment, and supplies.

Once these costs have been identified, the total cost can be divided by the number of claims processed to calculate the CPC. For example, if a healthcare organization spends $50,000 on cloud resources and $60,000 on other costs to process 4,600 insurance claims, the CPC would be $23.91 per claim.

Figure 12.4 – Q1 CPC

By tracking the CPC over time, healthcare organizations can also identify trends and make informed decisions about where to allocate resources to improve the financial performance of their healthcare services.

Revenue per claim

Revenue per claim (**RPC**) is a KPI used to measure the revenue generated from processing healthcare insurance claims in a healthcare organization.

To calculate the RPC, healthcare organizations need to identify all the revenue generated from processing healthcare insurance claims, including revenue from insurance reimbursements, copayments, and other fees.

Once the total revenue has been identified, it can be divided by the number of claims processed to calculate the RPC. For example, if a healthcare organization generates $990,000 in revenue from insurance reimbursements, copayments, and other fees for 4,600 insurance claims, the RPC would be $215.21 per claim.

Figure 12.5 – Q1 RPC

By tracking the RPC over time, healthcare organizations can also identify trends and make informed decisions about which insurance providers to prioritize and how to optimize their pricing strategy to improve the financial performance of their healthcare services.

Let's look at the activity-based cost model next.

Activity-based cost model

Activity-based costing (ABC) is a model used in cloud FinOps to allocate costs to the specific activities that cause those costs to be incurred. ABC aims to provide a more accurate picture of the true costs of delivering cloud services by considering all the costs involved in each activity, rather than just looking at the overall costs of a product or service.

In the context of cloud FinOps, the ABC model helps teams identify the specific activities that drive the cost of delivering cloud services, and how changes in those activities can impact the costs. This information is used to inform decisions about pricing, cost optimization, and resource allocation. For example, if a particular activity is found to be driving a large portion of the costs, FinOps teams can examine ways to improve the efficiency of that activity or find alternative methods of delivering the same service that may be less expensive.

Overall, the ABC model is a useful tool for cloud FinOps teams as it provides a more comprehensive and accurate understanding of the costs of delivering cloud services. This information can be used to drive more informed decision-making, optimize costs, and improve the financial performance of the business.

ABC and unit economics are both important tools used in cloud FinOps to help businesses understand the costs and revenue associated with delivering cloud services. Unit economics focuses on the financial performance of individual units or products, while ABC provides a detailed understanding of the costs associated with specific activities involved in delivering those units.

In the context of cloud FinOps, both unit economics and ABC are used to inform decisions about pricing, cost optimization, and resource allocation. By understanding the unit economics of cloud services, FinOps teams can make informed decisions about pricing, while the ABC model provides a more detailed understanding of the costs involved in each activity. This information can then be used to drive cost optimization and improve the financial performance of the business.

Together, unit economics and ABC provide a comprehensive understanding of the costs and revenue involved in delivering cloud services. Unit economics provides an overview of the financial performance of individual units, while ABC provides a detailed understanding of the costs associated with the specific activities that drive those costs. This information is essential for making informed decisions about pricing, cost optimization, and resource allocation, and for improving the overall financial performance of the business.

Let's summarize what we learned in this chapter.

Summary

In summary, unit economics is a framework for analyzing the financial performance of a business at the level of each unit or transaction. By examining the cost and revenue associated with each unit, businesses can gain insights into the profitability of their products or services and identify areas for improvement.

Some common metrics used in unit economics include customer acquisition cost, lifetime value, gross margin, and cost per unit. These metrics can be used to identify which products or services are most profitable, which customer segments are the most valuable, and where costs can be reduced. Effective unit economics analysis requires accurate data collection and analysis, as well as a deep understanding of the underlying business model. It is also important to consider external factors such as market conditions and competition when interpreting the results of unit economics analysis.

Overall, unit economics is a powerful tool for businesses looking to optimize their financial performance and achieve long-term success. By understanding the financial drivers of their business, organizations can make informed decisions about where to invest resources, which products or services to prioritize, and how to optimize their pricing and cost structure.

Let's look at the third and final case study of the book – implementing metric-driven cost optimization – next.

13

Case Study – Implementing Metric-Driven Cost Optimization and Unit Economics

This case study explores the challenges of allocating costs for containers, and the solutions we implemented to overcome these obstacles. We utilized a metric-driven decision-making process to determine when to purchase reservations or initiate the usage optimization process with our engineering team. Additionally, we calculated IT costs per unit of service to help us better understand the profitability of our DeliverNow platform.

In this case study, you will learn about the following:

- Our approach to cost allocation for containers and shared services in **Azure Kubernetes Service (AKS)**
- The metrics we implemented to assist with reservation purchases
- The unit metrics for the DeliverNow business requested by the finance department to calculate per-unit cost and profitability

Case study – Peopledrift Inc., a healthcare company

As a healthcare company, we specialize in providing IT services to hospitals and doctors' offices worldwide. Our highly-rated **online appointment scheduling (OAS)** platform is the industry leader in its space. Additionally, we acquired a small transport and logistics company three years ago, which is now the driving force behind our DeliverNow online platform. This platform enables us to deliver vaccines with speed, accuracy, and scalability to hospitals, pharmacy stores, and doctors' offices.

Last year, we made a strategic decision to exit the data center business and chose Microsoft Azure as our primary cloud for all workloads. We successfully migrated 150 internal and customer-facing applications, as well as 90 relational and non-relational databases. As our business expanded into new territories, the engineering team needed a significantly larger cloud capacity, which has resulted in a continuous need for provisioning infrastructure in Azure.

Challenges

We standardized AKS as our container orchestration platform. While allocating costs for app services, databases, and virtual machines using tags was straightforward, we encountered challenges in allocating costs for the AKS service. The two main issues were out-of-cluster costs and shared services. AKS represented approximately 40% of our cloud costs, and we had no effective method for allocating those costs accurately. We faced similar challenges with reservation purchases, using a schedule-based approach that led to purchasing reservations too early or too late, resulting in a 10% loss of savings. Finance requested we provide IT costs per unit of service, but our engineering team struggled to provide the necessary data due to a lack of unit economics metrics and KPIs.

Objectives

Peopledrift Healthcare has identified the following key objectives for optimization:

- A standardized approach for allocating costs of containers in AKS

- Allocate costs for in-cluster, out-of-cluster, and shared resources

- Establish metrics to aid reservation purchases and determine the optimal timing with an option to automate the purchase

- Create unit economics metrics for the DeliverNow platform

Solution

We decided to run our workloads in containers and opted for AKS as our container orchestration platform. However, AKS currently does not support cost allocation at the namespace, container, and cluster levels, which led us to explore various third-party cost allocation solutions. After considering the open source tools, we selected Kubecost. Implementing Kubecost was a seamless process, as it worked out of the box. We began allocating costs at the namespace level, which aligned with our unit of work for the DeliverNow platform's frontend and APIs.

Name	CPU Cost	RAM Cost	PV Cost	Network Cost	Shared Cost	Total
_ _ Idle _ _	$10,789	$22,345	$45,678	$11,234	$18,345	$108,391
DeliverNow_Frontend	$12,345	$26,789	$54,567	$13,456	$22,345	$129,502
Package_Billing_API	$8,123	$13,456	$26,789	$6,789	$11,567	$66,724
Package_Tracking_API	$15,678	$31,234	$63,456	$15,678	$25,345	$151,391
B2B_Integration_API	$6,789	$11,234	$22,345	$5,678	$10,123	$56,169
Streaming_Analytics_API	$9,567	$15,678	$31,234	$7,890	$13,345	$77,714
Grand Total Cost						$589,891

Figure 13.1 – DeliverNow AKS cluster cost allocation using Kubecost

By utilizing Kubecost, we were able to obtain our initial insight regarding the expenses incurred by our idle cluster. Having quantified these idle costs, our engineering team was then able to optimize resource allocation to appropriately size the cluster, resulting in idle costs being reduced to less than $1,000. Kubecost's effective cost analysis and resource allocation capabilities were key to allowing us to identify and minimize these idle costs.

We calculated several unit economics metrics for our DeliverNow platform to identify opportunities to optimize spending, reduce delivery time, and increase capacity without additional infrastructure costs:

- **Cost per vaccine delivery**: This metric calculates the cost per vaccine delivered to a hospital, pharmacy, or doctor's office. It considers the cost of transportation, storage, IT, and other associated expenses.

- **Utilization rate**: This metric measures the percentage of time for which the vaccine delivery platform is being used. It helps to identify areas of inefficiency in the system and can inform decisions about resource allocation.

- **Delivery time**: This metric measures the time it takes to deliver vaccines from the warehouse to the healthcare provider. It helps to identify areas where the delivery process can be optimized and streamlined.

- **Customer satisfaction**: This metric measures the satisfaction of healthcare providers with the vaccine delivery platform. It helps to identify areas where improvements can be made to increase customer satisfaction and loyalty.

By combining Kubecost data with Unit Economics metrics, we were able to identify that the business could benefit from adjusting the price of vaccine delivery for third-party partners. After examining the B2B integration IT cost, we found that our current pricing structure was not sufficient to cover the delivery expenses, resulting in an annual loss of approximately $100,000. By revising the pricing to reach a break-even point for delivery costs, we estimated that we could save up to $50,000 annually, while still maintaining a competitive price point for our third-party partners. This analytical approach, enabled by the utilization of Kubecost's data and Unit Economics metrics, facilitated an informed and data-driven pricing decision, resulting in significant cost savings for the business.

To purchase reservations in a timely and cost-effective manner, we adopted a metric-driven cost optimization approach. In the past, our engineering team purchased reservations on a quarterly basis, which sometimes led to suboptimal purchasing decisions and reduced overall savings. With the new approach, our FinOps team created a Power BI report using `RI_SAVINGS` data from Azure Cost Management and established a dashboard with a threshold value. Once the savings amount exceeded the threshold value, the FinOps team initiated the reservation purchase flow. To further streamline the process, we also implemented an automated reservation purchase system using Microsoft Flow, which enabled us to purchase 50% of our reservations in near-real time.

Next, let's investigate the benefits we gain by implementing container cost allocation, metric-driven cost optimization, and calculating unit economics metrics.

Benefits

The benefits we received by practicing the **Operate** phase of FinOps practice are as follows:

- Defined measurable objectives and goals for cloud cost management, such as reducing costs, optimizing resource usage, and improving cost visibility.

- Built a team of stakeholders from different departments, including finance, operations, and IT, to collaborate on FinOps initiatives. This team worked together to identify cost drivers, implement cost-saving measures, and monitor cloud costs.

- Implemented FinOps tools and software to automate cost management tasks, track spending, and optimize cloud usage. This helped us make informed decisions and identify cost-saving opportunities.

- Continuously reviewed and improved FinOps practices to ensure that they aligned with business goals and delivered value. We regularly analyzed cloud usage and costs to identify areas for optimization and cost reduction.

- Allocating costs for containers helped us understand our cloud usage and expenses more granularly. By tracking costs for each container, teams have identified which containers are driving the most costs and can take appropriate actions to optimize their resource usage.

- Allocating costs for containers ensured that teams are billed accurately for their cloud usage, allowing for more precise cost allocation among different projects, teams, and departments.

- Teams can identify cost-saving opportunities, such as rightsizing containers, optimizing container utilization, and identifying underutilized resources. This leads to overall cost savings.

- Metric-driven cost optimization enabled us to make informed decisions based on data and metrics rather than assumptions.

- Promoted a culture of continuous optimization and improvement. By regularly tracking and analyzing metrics, organizations can identify trends and patterns in their cloud usage and costs.

- By using unit economics, we gained a better understanding of the costs associated with our DeliverNow platform's business activities or transactions.

- Unit economics helped to ensure that cloud spending is aligned with business objectives, leading to better business outcomes and increased competitiveness in the market.

Let's summarize what we have learned so far from the case study.

Summary

In this case study, we examined Peopledrift healthcare's third phase of transformation, which is **Operate**. Our organization has successfully established a FinOps culture, which has enabled teams to optimize our cloud costs and drive business value. We achieved this through the implementation of container cost allocation practices, the utilization of metric-driven cost optimization strategies, and the adoption of unit economics. Our approach to cost management ensures that we only act on cost-saving measures when triggered by relevant data and that our cloud spending is aligned with business objectives to achieve better outcomes.

Congratulations on completing *FinOps Handbook for Microsoft Azure*! You have now gained a comprehensive understanding of how to optimize costs, increase efficiency, and manage budgets in the cloud. With the knowledge and skills acquired from this book, you will be able to make informed decisions that align with your business goals, reduce unnecessary expenses, and ultimately, maximize your return on investment. We hope this book has been a valuable resource for you and wish you the best of luck in your future endeavors with Azure Cloud FinOps.

Index

A

ABC allocation 13
  activity rate 14
  advantages 13
  cost pool 13
  example 13, 14
  measure 14
  overhead 13
  product 13
ACI cost allocation 173, 174
activity-based cost (ABC) model 13, 215
AKS clusters
  abandoned workloads, remedying 188, 189
  container requests right-sizing 187, 188
  cost optimization 184
  local disks, resizing 186
  persistent volumes right-sizing 189, 190
  reserved instances 186, 187
  underutilized nodes, managing 184, 185
AKS cost allocation 175-180
alerts
  creating, in Azure cost analysis 38

anomaly alerts
  spending 40, 41
Apptio Cloudability 168
  URL 169
automated budget
  actions 162-168
automated tag
  compliance 153-156
  governance 153-156
  inheritance 153-156
automated VM
  shutdown and startup 156-161
Azure 28
  commitment-based rate discounts 98
  on-demand and elastic nature 28
Azure Advisor 5
  accessing, with CLI 78, 79
  accessing, with portal 77, 78
  recommendations, for reservations 105-107
Azure Advisor, recommendations
  providing, for usage optimization 76
  reference link 77

Azure App Service 92

Azure CLI 4

Azure Consumption APIs 45

Azure Container Instances (ACI) 171

Azure Content Delivery Network (CDN) 17

Azure cost analysis
  alerts, creating 38
  budget, creating 33
  budget, managing 33

Azure Cost Management + Billing 5

Azure Cost Management (ACM) 144

Azure Cost Management
    (ACM) Power BI app
  reservations, purchasing scenario 102-105
  using 100-102

Azure Hybrid Benefit
  enabling, for Linux VMs 84, 85
  enabling, for managed instances 87, 88
  enabling, for SQL databases 87, 88
  enabling, for SQL VMs 87, 88
  enabling, for Windows VMs 84, 85

Azure Hybrid Benefit FAQ 75

Azure Kubernetes Service (AKS) 93, 94, 171

Azure Monitor 6

Azure portal 45
  cost analysis 18, 19

Azure Pricing Calculator 6

Azure Spot market 122
  Spot VM 123
  VM discounts, estimating 122, 123
  VM Scale Sets 123-127

Azure usage data
  obtaining 45

Azure WAF 7
  pillars 7

B

baseline
  creating, with WAF Cost Optimization
    assessment 7-13

budget 33
  creating, in Azure cost analysis 33
  managing, in Azure cost analysis 33
  spend, tracking 38

budget alerts 38
  setting up 39, 40

C

Cast.ai 169
  URL 169

Center of Excellence (CoE)
  establishing, for cloud cost management 152

CLI
  used, for accessing Azure Advisor 78, 79

cloud cost management
  CoE, establishing 152

CloudHealth by VMware 169
  URL 169

cloud unit economics 208
  benefits 208

commitment-based rate discounts 98
  in Azure 98

commitment-based rate discounts, in Azure
  Microsoft Azure Consumption
    Commitment (MACC) 99
  Microsoft Enterprise Agreement 98

containerized workload
  FinOps challenges 172

Continuous Export 45

cost **73**
  versus performance 95
  versus reliability 95
  versus security 95
cost allocation
  from accounting point of view 13
cost allocation, Azure for FinOps 15
  account, using 15
  management group, using 16
  resources tags, using 17, 18
  subscriptions hierarchy, using 16
cost analysis, Azure portal
  accumulated and forecasted cost 20, 21
  cost grouped, by management group 22-24
  cost grouped, by service 21, 22
  cost grouped, by tag 24
  custom cost analysis views, creating 24, 25
  custom cost analysis views, saving 24, 25
  custom cost analysis views, sharing 24, 25
  exploring 18
  offer type, identifying 19, 20
cost driver-based forecasting 51
Cost Management automation
  reference link 45
Cost Management connector
  setting up, in Power BI 46-50
Cost of Goods Sold (COGS) 45
cost optimization
  business case, writing 136
  Orion business analytics platform
    cost optimization 136
cost per claim (CPC) 213
  calculating 213
cost per patient encounter (CPE) 210
  calculating 210, 211
costs back, to business benefits
  tracking 209, 210
custom Azure Workbook
  used, for usage optimization targets 80

D

Daily Active Users (DAU) 64
Database Transaction Unit (DTU) 86
direct cost metrics 209
  versus indirect cost metrics 209

E

EA Portal 45
Electronic Health Record System (EHR) 29
engineering teams
  motivating, to take actions 152
engineering teams, actions
  incentivizing 153
  penalizing 153
eviction 127
  handling, best practices 130, 131
exponential smoothing 55

F

FFmpeg utility 130
FinOps challenges
  for containerized workload 172
forecasting 44, 45
  by application 55
  fully loaded cost 58-61
  usage charges, identifying by
    application 55-58
**forecasting, based on manual
  estimates 50, 51**
forecasting, based on past usage 51-55
**forecasting models, describing
  in Power View**
  reference link 55

G

generally available (GA) 133
gross margin per patient
        encounter (GMPE) 212
    calculating 212, 213

I

indirect cost metrics 209
    versus direct cost metrics 209
infrastructure as code (IaC) solution 18
interruptible workload
    architecture pattern 131
Iron Triangle 72

K

key performance indicators (KPIs) 28, 208
    Azure Hybrid Benefit utilization 75
    cost avoidance of unattached disks 74
    defining 29, 30
    defining, attributes 29
    developing, for consistent reporting 28
    measuring 30
    need for 29
    reporting 30
    setting 73
    storage account tiers 76
    tagging by business units 74
Kubecost 174
    reference link 174

L

lagging KPIs 28, 29
leading KPIs 28
Linux VMs
    Azure Hybrid Benefit, enabling 84, 85

M

marketing development budget
    creating 37
marketing production budget
    creating 37
MDCO triggers
    reservation purchases, automating
        based on 202-206
Metric-Driven Cost Optimization
        (MDCO) 191
    principles 192, 193
    reservation reporting, purchasing with
        Power BI based on 193, 194
metrics, for unit economics
    cost per claim (CPC) 213
    cost per patient encounter (CPE) 210, 211
    developing 210
    gross margin per patient encounter
        (GMPE) 212, 213
    revenue per claim (RPC) 214
    revenue per patient encounter
        (RPE) 211, 212
Microsoft Azure Consumption
        Commitment (MACC) 99
    lifecycle 99
Microsoft Customer Agreement
        (MCA) billing hierarchy 16
Microsoft Enterprise Agreement 98
    benefits 98

O

Objectives and Key Results
        (OKRs) 71, 73, 142
    examples 73, 74
    setting 73

Online Appointment Scheduling
    (OAS) **63, 142 , 217**
Operate phase 220
operational KPIs 28
Orion business analytics platform
    cost optimization 136-139
overall Marketing department budget
    creating 37

P

Peopledrift Healthcare 142
    challenges 142
    objectives, for optimization 142
    Optimize phase of FinOps
        practice, benefits 146
    solutions 142, 143
Peopledrift Healthcare, case study 63, 64
    benefits 66, 67
    challenges 64
    objectives 64
    solution 64, 65
Peopledrift Healthcare, solutions
    10% PAYG v-team 145, 146
    RI savings v-team 144
    SP v-team 145
    zero-waste v-team 144
Peopledrift Inc. healthcare company
    case study 217, 218
Peopledrift Inc. healthcare
    company, case study
    benefits 220, 221
    challenges 218
    objectives 218
    solution 218-220

performance
    versus cost 95
portal
    used, for accessing Azure Advisor 77, 78
Power BI
    Cost Management connector,
        setting up 46-50
    used, for purchasing reservation reporting
        based on MDCO 193, 194
Power BI Desktop 4
production marketing website budget
    creating 34-37
project management triangle 72
    cost 73
    for goal setting 73
    quality 73
    time 73
project management triangle, method
    for goal setting 72

R

rate optimization 97
Recovery Point Objectives (RPOs) 95
Recovery Time Objectives (RTOs) 95
Red Hat Enterprise Linux (RHEL) 84
reliability 95
    versus cost 95
research and development (R&D) 209
reservation opportunities, for workload
    identifying 100
reservation purchases
    automating, based on MDCO
        triggers 202-206

reservations
    auto-renewal 114
    Azure Advisor recommendations 105-107
    cadence 107-112
    cancel (return) 119
    chargeback report 114-116
    details 112, 113
    exchange 116-119
    purchase 107-111
    savings 114-116
    thresholds, setting for purchasing 194-201
Reserved Instance Planner 169
revenue per claim (RPC) 214
    calculating 214
revenue per patient encounter (RPE) 211
    calculating 211, 212

S

savings plans
    discounting strategies 133, 134
    purchasing 135, 136
    reference link 134
    versus reserved instances 134
scorecards 31
    examples 31, 32
security
    versus cost 95
Selling, general, and administrative
        (SG&A) costs 13
service-level agreement (SLA) 95, 127
shared AKS clusters 180-184
Spot Priority Mix 132, 133
Spot VM
    caveats 127
    eviction rate details 129, 130
    pricing history 128-130

Spot VM, caveats
    eviction type and policy 127
    limitations 127
Stock Keeping Unit (SKU) 86
storage accounts
    upgrading, to General-purpose V2 88, 89
strategic KPIs 28
SUSE Linux Enterprise Server (SLES) 84

T

team's performance
    benchmarking 31-33
third-party FinOps tools 168
    Apptio Cloudability 168
    Cast.ai 169
    CloudHealth by VMware 169
Triple constraints 72

U

usage optimization 80
    Azure Advisor recommendations 76
usage optimization targets, with
        custom Azure Workbook 80
    Azure App Service 92
    Azure Hybrid Benefit, enabling
        for Linux VMs 84, 85
    Azure Hybrid Benefit, enabling for
        managed instances 87, 88
    Azure Hybrid Benefit, enabling
        for SQL databases 87, 88
    Azure Hybrid Benefit, enabling
        for SQL VMs 87, 88
    Azure Hybrid Benefit, enabling
        for Windows VMs 84, 85
    Azure Kubernetes Service (AKS) 93, 94

right-sizing, underutilized SQL databases 86

right-sizing, underutilized
    virtual machines 82, 83

storage accounts, upgrading to
    General-purpose V2 88, 89

tagging 80-82

unattached discs, deleting 89, 90

unattached public IPs, deleting 91

V

vacancy 139

virtual machines (VMs) 141

virtual team (v-team) 143

VM Scale Set
    creating 124-127

W

WAF Cost Optimization assessment
    used, for creating baseline 7-13

Windows VMs
    Azure Hybrid Benefit, enabling 84, 85

`packtpub.com`

Subscribe to our online digital library for full access to over 7,000 books and videos, as well as industry leading tools to help you plan your personal development and advance your career. For more information, please visit our website.

Why subscribe?

- Spend less time learning and more time coding with practical eBooks and Videos from over 4,000 industry professionals

- Improve your learning with Skill Plans built especially for you

- Get a free eBook or video every month

- Fully searchable for easy access to vital information

- Copy and paste, print, and bookmark content

Did you know that Packt offers eBook versions of every book published, with PDF and ePub files available? You can upgrade to the eBook version at `packtpub.com` and as a print book customer, you are entitled to a discount on the eBook copy. Get in touch with us at `customercare@packtpub.com` for more details.

At `www.packtpub.com`, you can also read a collection of free technical articles, sign up for a range of free newsletters, and receive exclusive discounts and offers on Packt books and eBooks.

Other Books You May Enjoy

If you enjoyed this book, you may be interested in these other books by Packt:

AWS FinOps Simplified

Peter Chung

ISBN: 978-1-80324-723-6

- Use AWS services to monitor and govern your cost, usage, and spend
- Implement automation to streamline cost optimization operations
- Design the best architecture that fits your workload and optimizes on data transfer
- Optimize costs by maximizing efficiency with elasticity strategies
- Implement cost optimization levers to save on compute and storage costs
- Bring value to your organization by identifying strategies to create and govern cost metrics

Industrializing Financial Services with DevOps

Spyridon Maniotis

ISBN: 978-1-80461-434-1

- Understand how a firm's corporate strategy can be translated to a DevOps enterprise evolution
- Enable the pillars of a complete DevOps 360° operating model
- Adopt DevOps at scale and at relevance in a multi-speed context
- Implement proven DevOps practices that large incumbents banks follow
- Discover core DevOps capabilities that foster the enterprise evolution
- Set up DevOps CoEs, platform teams, and SRE teams

Packt is searching for authors like you

If you're interested in becoming an author for Packt, please visit `authors.packtpub.com` and apply today. We have worked with thousands of developers and tech professionals, just like you, to help them share their insight with the global tech community. You can make a general application, apply for a specific hot topic that we are recruiting an author for, or submit your own idea.

Share Your Thoughts

Now you've finished *FinOps Handbook for Microsoft Azure*, we'd love to hear your thoughts! Scan the QR code below to go straight to the Amazon review page for this book and share your feedback or leave a review on the site that you purchased it from.

`https://packt.link/r/1-801-81016-8`

Your review is important to us and the tech community and will help us make sure we're delivering excellent quality content.

Download a free PDF copy of this book

Thanks for purchasing this book!

Do you like to read on the go but are unable to carry your print books everywhere? Is your eBook purchase not compatible with the device of your choice?

Don't worry, now with every Packt book you get a DRM-free PDF version of that book at no cost.

Read anywhere, any place, on any device. Search, copy, and paste code from your favorite technical books directly into your application.

The perks don't stop there, you can get exclusive access to discounts, newsletters, and great free content in your inbox daily

Follow these simple steps to get the benefits:

1. Scan the QR code or visit the link below

https://packt.link/free-ebook/9781801810166

2. Submit your proof of purchase
3. That's it! We'll send your free PDF and other benefits to your email directly